Should We Colonize Other Planets?

New Human Frontiers series

Nigel M. de S. Cameron, *Will Robots Take Your Job?*
Harry Collins, *Are We All Scientific Experts Now?*
Everett Carl Dolman, *Can Science End War?*
Mike Hulme, *Can Science Fix Climate Change?*
Sheila Jasanoff, *Can Science Make Sense of Life?*
Margaret Lock & Gisli Palsson,
Can Science Resolve the Nature/Nurture Debate?
Hugh Pennington, *Have Bacteria Won?*
Hilary Rose & Steven Rose, *Can Neuroscience
Change Our Minds?*

Should We Colonize Other Planets?

ADAM MORTON

polity

First published in 2018 by Polity Press

Polity Press
65 Bridge Street
Cambridge CB2 1UR, UK

Polity Press
101 Station Landing
Suite 300
Medford, MA 02155, USA

ISBN-13: 978-1-5095-2511-9
ISBN-13: 978-1-5095-2512-6 (pb)

A catalogue record for this book is available from the British Library.

Typeset in 11 on 15 pt Adobe Garamond
by Toppan Best-set Premedia Limited
Printed and bound in Great Britain by CPI Group (UK) Ltd, Croydon

The publisher has used its best endeavours to ensure that the URLs for external websites referred to in this book are correct and active at the time of going to press. However, the publisher has no responsibility for the websites and can make no guarantee that a site will remain live or that the content is or will remain appropriate.

Every effort has been made to trace all copyright holders, but if any have been inadvertently overlooked the publisher will be pleased to include any necessary credits in any subsequent reprint or edition.

For further information on Polity, visit our website:
politybooks.com

CONTENTS

Acknowledgements   vi

1 Escape from Earth?   1

2 The Colony Solution   22

3 Problems with Colonies   40

4 Costs of Colonization   57

5 Colonization without Humans   78

Conclusion: Why Human Colonization
      is a Bad Idea   99

Further Reading   103
Index   121

v

In writing this book I have had helpful suggestions from and discussions with a number of people. I should mention Susanna Braund, Richard Johns, Toph Marshall, Lije Milgram, Pascal Porcheron, Mark Trischuk, and two anonymous readers from Polity. Of course I have not taken all their advice or answered all their objections.

ONE

Escape from Earth?

The aim of this book is to look critically at some of the arguments for thinking that human beings should travel and colonize beyond the earth. Why should we want to leave this beautiful planet to which we are so well adapted? Some of these arguments suggest escapes from dangers facing the human species, and others are driven by scientific curiosity and the love of adventure. This chapter describes the dangers, chapter 2 the proposed solutions, chapter 3 problems with these solutions, chapter 4 their monetary and opportunity costs, and chapter 5 alternatives to the standard colonization proposals. At the end there is a very short conclusion drawing some general morals. I argue that the dangers to a colony are greater than we often suppose, that the costs are likely to be higher, and that we are ignoring serious alternatives and possibilities. Moreover, there are hard questions about the desirability of colonization. The first topic – dangers and costs – is a matter of finding the right facts. This is not always easy, but it is common in discussions of topics like this. The second topic – which alternatives we want – is a matter of making readers confront what on reflection they value. These are moral arguments and work by setting up suitable reflection. They are not common in this kind of topic, and it is a novel aspect of this book to give them equal weight. We begin with the dangers. What are they?

Earth is not heaven. Many things go badly on our planet: some of these are old, by earthly standards, some are new, some are less than they used to be, and some we are making worse. The Four Horsemen of the Apocalypse – traditionally Conquest, War, Famine, and Death, though more modern names are enslavement/ inequality, war, environmental damage, and murderous hatred – are still on the horizon and have the capacity to extinguish the human species. War and environmental damage are the species killers, but they are abetted by inequality and hatred. Inequality and famine promote hatred and raise the risk of war. (We have dodged nuclear war for seventy years, for reasons that no longer seem clear, but have substituted a large number of terrible but less catastrophic conflicts.)* The number of nuclear

---

* References are in the further reading section at the end of the book. There is a fair amount of additional information in this section, so I encourage the reader to browse through it.

nations has grown, and some of them have fanatical and irresponsible rulers. The dangers to human life are not confined to the destruction from the bombs themselves: fallout, destruction of crops, and nuclear winter would kill far more, possibly all of us.

Hatred can obviously lead to war. But private hatred can be large-scale also. The greatest danger exploits biotechnology. Viruses can be designed, manufactured, and targeted to order these days. A variant of the Ebola virus, for example, that spread not just via blood and other fluids but also through the air, perhaps on people's breath, would have catastrophic possibilities. Most viruses have reservoirs in animal populations from which they can mutate and find ways of infecting humans. With greater density of human population, diseases spread more readily; we use more intensive farming methods, and in some parts of the world we hunt and eat animals to which we did not pay much attention before. People travel more readily in the contemporary world, and air travel is so rapid that a person can get on a flight showing no signs of a disease and be highly infectious when they arrive at their destination. The great danger is of a mutation in a virus with the terrible combination of three factors. It would come to be transmitted readily from person to person; it would become more deadly by finding a chink in human immune defences; and it would have an incubation period between infection and presentation of symptoms which allowed it to spread by air travel around the globe before governments realized what was going on. Such a disease could take an enormous toll of human life and might set civilization back for centuries. It could threaten the survival of the species. It could happen as a result of bad luck plus modern conditions of life. But it could also happen by deliberate design, through the targeting of some nation or ethnic group, and then spread to cover the globe and all of humanity.

These are the evils that our species brings upon itself. There are also evils that come from outside. The threat most often mentioned is an asteroid strike. It has features in common with other catastrophes, so I will give it perhaps disproportionate attention. Asteroids are big chunks of rock which are numerous in the outer reaches of the solar system and occasionally wander inwards, where they can strike planets. The surfaces of other rocky planets are pitted with craters from the impact of asteroids. They are also visible on Earth to those who know what to look for, showing past asteroid strikes on this planet, but they are less obvious because of oceans, vegetation, and the constant churning of the earth's surface associated with continental drift.

The best-known asteroid impact is the one that many believe to have played a role in the extinction of the dinosaurs (except for birds). The theory is that, around 66 million years ago, an asteroid as large as 15 km across struck the earth at the coast of Yucatán. It would have created a worldwide cloud that lasted a year and was thick enough to inhibit photosynthesis, thus depriving animals of food and extinguishing many species. Fate is chancy: an impact in the deep ocean which covers most of the earth, or in less sulphur-rich rock, might have been less devastating. If something similar happened tomorrow, the survival of the human species would be far from guaranteed.

There are many other cases; asteroids strike Earth regularly. Most of them are small and burn up in the earth's atmosphere. If they reach the surface they do minimal damage. But not all of them are like this. There is an impact with the force of a small atomic bomb every ten years or so. Luckily most of these are dissipated in the upper atmosphere. But if an impact of a size that happens every few centuries struck a populated area the result would be a catastrophe. It would be much more serious than most wars or terrorist incidents. The explosive impact itself is only one of the sources of destruction. There would likely be effects on the earth's climate for a good while. We are lucky to have an atmosphere that mitigates the danger, but it is not enough to make us safe.

There have been a variety of proposals for reducing or eliminating the danger of a catastrophic asteroid strike. We can often detect asteroids when they are very distant and predict their danger to Earth. With advances in telescopes, particularly telescopes orbiting the earth – significantly for our purposes, telescopes located on the moon or at the gravitationally stable Lagrange points – we can increase our warning time. As an asteroid approaches, we can refine our predictions and eventually foretell where on Earth it should strike. We could then evacuate vulnerable areas of the planet. The amount of warning we have is crucial.

A more active response would be to deflect or destroy the asteroid. Some of these responses are rather science-fictional, involving astronauts on rockets and nuclear explosions. Deflection of the asteroid from its fatal course could be managed in a number of ways. We might exploit the gravitational tug of an object made to orbit the asteroid, or we might play cosmic billiards using a different asteroid or other body directed towards the one headed towards us. It has even been suggested that we could focus light from the sun at the asteroid, producing a subtle pressure that, given enough time, would change its course. NASA has announced plans for a fairly simple deflection strategy, the DART plan (Double Asteroid Redirection Test!), which involves impacting the asteroid with a spacecraft whose momentum would change the asteroid's trajectory. The deflection strategies typically require more advance knowledge than the fragmentation strategies, which are risky and fall into the category of last resort. No one has begun to construct the hardware that any of these require. But a lot of effort from many astronomers and others is going into locating and predicting "heavenly" bodies that might be threats.

There are other dangers in the earth's cosmic environment. Cosmic rays from outer space are variable, and we do not know all the sources of their variation. The movement of the solar system through the galaxy may be one source. Supernovas – exploding stars at the end of their lives – could have bad effects by destroying the earth's ozone layer with X-rays and gamma rays, thus allowing ultraviolet light from the sun to damage life on Earth. And, for reasons that are still very obscure, there seems to be a mass extinction of life on Earth every 27 million years or so. Supervolcanoes may be one explanation and volcanic activity remains a major threat. It may be as great a threat as asteroids, though it does not get as much discussion. We do not now have the capacity to mitigate any of these further dangers. As science and technology progress, we may come to understand and even to take protective measures against some of these. But not immediately, and perhaps not for centuries.

All these hazards can interact and amplify one another's effects. Some of them are very likely to do this. We are now seeing the first effects of global warming. It will become worse in the coming years and poses a great threat to human life. The resulting climate changes make war and disease more likely, as does population increase. A world with many refugees is likely to be one with even greater inequality and will have many menacing sources of disease. Public health systems are less of a barrier to the spread of disease when their work is made more difficult by inequality and poverty. The agriculture that feeds us is threatened by climate change and the effects of pollution from the industry and industrial farming necessary to support a large population. So, one nightmare scenario involves a mutated virus attacking a population that is forced to spread throughout the globe because of climate change, leading to a worldwide pandemic, which is compounded by food supply failures, which also increase the death toll of the disease, in a feedback spiral. Such a spiral could also be precipitated by an asteroid strike on a world already affected by global warming and its social disruption. In fact, all of the major hazards are likely to compound one another's effects. We must remember that the organization of human life was largely unchanged for hundreds of thousands of years and has only recently emerged into a time of relative but uneven health and prosperity for many people, from which it could retreat fairly easily. The Horsemen like to ride together: the combination of war, environmental damage, and hatred, especially if aided by some calamity from outside, would be horribly and almost irresistibly powerful.

It is probable that, if we are complacent, sooner or later there will be a catastrophe, unprecedented in the brief time that the planet has harboured humans, perhaps ending our species. It is as if we are well along a series of coin tosses which on this planet have miraculously all landed heads. (Elsewhere intelligent life may not have been so lucky.)

The dangers are real. But they raise many further questions for later chapters. Among them are:

Do these probabilities come with a definite timeframe? After all, a run of heads can go on indefinitely.

How much do the dangers depend on present-day politics and society, and are there reasonable political changes that can affect them?

One way out has been suggested more and more often in recent years: we become less tied to this planet and its fate. In the long run, the idea is that we colonize other planets, in this solar system or eventually others, so that when human life on Earth comes to an end we have an attainable refuge beyond Earth where at least some people survive. (More than mere survival perhaps: some imagine that humans might become a galaxy-wide species.)

Shorter-term plans focus on Mars – inevitably, since Mars is the only remotely plausible candidate in the solar system without invoking technological capacities that are probably centuries away. These plans are largely organized by private corporations, such as the Dutch outfit Mars One and, more plausibly and responsibly, Elon Musk's SpaceX. (NASA has no immediate colonization plans involving humans.) Publicity for both of these invokes the rhetoric of human species survival. I describe these solutions in more detail in the following chapter. Again there are basic questions that we need to answer. They include:

Could such a colony succeed? What are the chances of such a project landing colonists on Mars and the colony surviving for any length of time?

Would a colony work as a refuge? If humanity dies out on Earth, what are the chances that it could survive elsewhere? Where? For how long? When might this be feasible? Might colonization aimed at our survival even increase the dangers for us on Earth?

At what cost? Would the probability of survival be accompanied by a decreased quality of life on Earth? What might it cost the environment? What benefits might we have to give up in exchange?

Dangers from machine intelligence

Worries about artificial intelligence have haunted recent discussions of planetary colonization. We rely on it and fear it. While we are developing an economy that depends increasingly on intelligent machines, at the same time we are becoming worried that, instead of serving us, the machines will control and perhaps even eliminate us. Profound thinkers such as Stephen Hawking and brilliant technological figures such as Elon Musk have expressed such worries. There is a tendency to pronouncement and appeals to the intuition of smart and experienced people here, as well as a lack of explicit argument. So the arguments I shall present are partly reconstruction. But they have a lot of power.

The perceived danger of advances in artificial intelligence and robotics comes from combining two real factors. The first is the unpredictability for us of machines designed by machines. Computers are even now both helping to design chips for other computers and debugging software. The development that many consider inevitable is that the overall design of computing machinery will become so complicated that most of it will also be entrusted to computers. Computers will design computers. But we already have only a very incomplete ability to predict what a computer will do on the way to solving a problem. One standard example is computer proof, most famously of the four-colour theorem, where a computation covers so many details of so many cases that a human mathematician cannot check them all but must trust that the machine is programmed correctly. Another standard example is a chess program that can beat human champions. The person or, more likely, team who wrote the program will be able to describe its overall aim and strategy but will usually be unable to predict what it will do at any particular move. It will check through a branching tree of possible positions from each move that it considers to a greater depth than any human being could keep in mind. And it will be sensitive to patterns that our nervous systems will not register. As a result, if a computer designs another computer, probably a more powerful one, the result will most likely have many features that were not anticipated by whoever designed and programmed the first one. The details of machine-designed machines will be beyond our control or understanding.

Adding to the unpredictability of programs is their literalness. If you tell an AI program "take everything out of the fridge and put it on the table," you may find that all the dirt in the fridge and the light bulb and the drawers are also on the table. It takes you at your literal word, but it is hard for humans to see what their literal words, especially taken in large quantities, actually entail. The result is unexpected consequences that paradoxically comply with what we asked for. In philosophy and computer science this is known as the frame problem.

These may not seem such terrible dangers. After all, the fact that we cannot grasp all the details of a computer proof or a chess program does not harm us. And there are benefits: we can get more mathematical results and challenge our chess-playing skills to a higher level. But the danger emerges when we combine unpredictability with Darwinian considerations.

Evolution combines two elements. The survival of the fittest says that individuals better suited to their environments live longer and have more offspring. And inheritance says that random changes in the design of individuals can be passed on to their offspring. Together, these mean that, when individuals compete for resources, the ones that compete best get more, and over time their descendants tend to crowd out the others with whom they compete and drive them to extinction.

At first sight the analogy between the biological case and the computer case seems to break down at several points. Computers consume electricity rather than human food and do not need oxygen or their bodies to be kept within such a narrow range of temperatures, so the competition is not direct. And there is nothing essential to the fact that one computer is designing others that makes it "want" there to be many of these others. Computers are not built to love their offspring.

But, if we say just this, we are failing to see how the unpredictability point and the Darwinian point combine. Since we cannot predict in detail what computer-designed computers will be like, it might happen that an unforeseen quirk in the design of some computer leads it to operate in such a way as to produce as many as it can of others like it or simply others that it has designed. In fact, we can argue that it is inevitable that such machines develop. For the unpredictability of complex programs is like a kind of random mutation, and once one machine set to create more like itself has come into existence it will make many others and will act to neutralize anything that reduces the number. It is presumably some such process that results in reproductive strategies in animals. So once a program exists that makes physical computers that run the same programs, and also takes measures that there be as many of them as possible, the number of machines of this general sort will increase, and they will act so as to reduce opposition to them. They might then compete with humans for energy and other resources. Thus a conflict is set up.

This is the best case I can make of an existential threat to humanity from intelligent machines. But now I must debunk it. Although the robot threat is often given as a reason why we should establish colonies beyond Earth, I have no doubt that the threat is exaggerated. I see three weak places, and one interestingly ambiguous topic, in the fear thus raised.

First, there is the distinction between hardware and software. Computers can design other computers, because their programs were designed by other programs that have been successful at designing programs for designing programs. Or they can actually build them. An automated computer-making factory could do both, and they could be closely integrated. But these can be kept apart. Computers can be unplugged;

simpler fallback software can be activated; alternative "checks and balances" software can be developed, to be put into use in times of crisis.

Second, we can test software before putting it into the physically real world. The most promising means would be specific programs that probe the recesses of proposed software in order to detect just this phenomenon. (They would be vaguely analogous to the immune system in human physiology.) We can create an artificial reality in which software can operate so that its long-term devious potentialities can emerge. The software we are investigating does not have to "know" that it is not operating in the real world. These measures are also far from infallible, but they do reduce the probabilities. Recent developments in integrating human and computer proofs of mathematical theorems may well play an important role in this.

Third, computers and biological organisms are not in direct competition. Biology needs food, water, and oxygen, while computers do not need much besides electricity. (This is a central reason why they are better suited than humans to operate in space.) So a program wanting to produce many descendants of itself will not care who is consuming the crops or metabolizing the air. It may want the lion's share of power generation and dominate the use of silicon and other materials, but that is the way that things are likely to develop anyway, whether computers are serving human interests or their own.

Remember that the computer villains in this conspiracy story are super-intelligent and well informed. They are extremely rational in the sense that, given enough information, they make decisions that are in their interest. From their point of view, their human great-grandparents are slow, dull-witted, and irrational. So they will not see humans offering a very threatening competition – no more so than other semi-intelligent life forms such as whales or parrots. Humans would be no more the target than many other species that exploit similar resources.

The rationality of these hypothetical intelligences contrasts with human shortsightedness and self-destruction. Humans often hate other humans even when it is in their interest not to do so. Humans regularly adopt strategies that they could easily see will damage what they value most. We find it hard to see our way to measures that are best for all of us. These are undeniable dangers to our species. Compared to climate change, nuclear war, and deliberate epidemics, the threat of all-powerful computers pales. Whatever dangers they pose, it is very little compared to the dangers we present to ourselves.

Discussions of the threat of artificial intelligence from AI professionals tend to be sceptical about exaggerations of the threat. This contrasts with the Hawking and Musk line, which as far as I know has never been worked out in detail in books or articles but gets a lot of publicity from brief quotations on the Internet. In a recent interview Musk proposes that institutions be set up to contain the threat. That is sensible and suggests planning rather than panic.

There are dangers humanity brings upon itself, dangers from beyond the earth, and threats from our own creation, artificial intelligence. All have been cited as motives for colonizing other planets. Each raises basic questions that must be answered. There are also questions that arise whichever threats we are considering. The most basic is how to decide what to do. Do we weigh benefits against probabilities, acknowledging that our estimates of both will be extremely rough? Do we allow sufficiently awful possible outcomes to veto options without calculating precise costs and probabilities? (An extreme example of this is the so-called precautionary principle.) Should the decisions be made by societies and governments or by entrepreneurs and private organizations? We do not need a complete and systematic attitude to these – just as well, since no one has one – but we must not wander blindly between them. We must be honest about how we are negotiating the questions, so that our methods as well as our facts are open to criticism.

Colonization versus spaceflight

My concern is with plans to colonize other planets. But we have to get there first, and before that we have to know what to expect we will find. Colonization is thus associated with spaceflight, with or without human crews. Spaceflight that is not directed at colonization is not my concern here, though spaceflight that prepares for or maintains a human colony obviously must be. Most spaceflight that does not ferry colonists is best done robotically. Most, essentially all, space exploration is already performed without human piloting. We – that is, NASA and the European Space Agency – have had a string of successes in sending precisely choreographed crewless spacecraft into the solar system: rovers on Mars, the Cassini spacecraft manoeuvring delicately around Saturn's rings and moons and sending the Huygens probe to Saturn's moon Titan, the New Horizons probe to Pluto and beyond, and the Rosetta spacecraft reaching a comet and sending its lander module Philae to it. These have needed a good degree of autonomy, given the time delay in getting signals from Earth to them. Nearer to home, Japanese and European robotic spacecraft have resupplied the international space station. A crucial factor behind the success of these ventures is the advance in computing power that can be included in probes and spacecraft. (The human missions to the moon in the years after 1969 were partly motivated by national prestige. But at that point in technological progress, on-the-spot human capacities may have been more necessary. One reason the Apollo expeditions have not been repeated may be that we can now gain just as much knowledge more cheaply without any risk to human life.)

There are few advantages to human crews in purely scientific spaceflight. To begin, human crew add a lot of mass. Humans are largish air-breathing mammals. Because they are mammals they have to keep their metabolism revved up all the time and have to keep their body temperature more or less constant. (Even hibernating bears have to accumulate considerable fat reserves to get through the winter.) So human passengers/pilots will need oxygen-generating equipment, water-recycling equipment, heating, and food.

It is conceivable that in the further future we will have mastered suspended animation for humans, but humans will remain large objects with much of their bodies specialized for tasks that are not needed for the purpose. Useless cargo: you really pay an excess baggage fee when there are humans involved.

TWO

The Colony Solution

Earth has a wide range of actual and potential troubles. Some of them threaten widespread environmental damage; some widespread human misery; some enormous loss of life; some the extinction of the human species; and some the extinction of life on Earth. I have listed these in order of likelihood: environmental damage is certain, the extinction of life on Earth is very unlikely within thousands of years, and the others are in between. We humans worry about our own species, so possible human extinction looms large for us. (Whether it should be the very worst-case scenario is something to which I shall return in chapter 5.) No one doubts the enormous value of anything that reduces the threat of nuclear or biological war, and anything that cures humanity of its hatreds would be magical. But it is hard to be optimistic, given the state of the world and given the lessons of history. This is the motive for a simple reaction: get out.

The premise is pessimistic but, some would say, probable enough to act on. If there is a sizeable threat of human extinction on Earth, then a human presence beyond Earth would give another chance for the species to survive. A colony on Mars is the only way of achieving such a presence within the limits of present technology, and indeed any technology that will be available in the foreseeable future. (Predicting technological progress is an occupation for the foolishly overconfident. A wild guess would be that it will be a couple of centuries before a colony anywhere besides Mars is an option.) In the next chapter I discuss problems with the idea. This chapter simply presents it. It is a short chapter because, although there is a lot of publicity at the moment about plans for Mars colonies, these lack detail, both about how a colony would be transported and set up and about how it would survive. A criticism is in some ways easier than an exposition, because one can lay out a range of possibilities and discuss the problems that each of them may run into. On the other hand, the existing plans have a certain element of solving problems on the fly and of trust in technological development. That is not necessarily a bad way to think, but it is rather frustrating for a responsible commentator. Very little in this chapter is original. It provides a factual basis for later chapters.

Why Mars?

Mars is the fourth planet from the sun, after Mercury, Venus, and Earth. Its distance from Earth varies between a theoretical minimum of 54.6 million km and a maximum of 401 million km, with an average separation of 225 million km (the distance varies because the earth gets around the sun more quickly, being nearer to it, and both planets have elliptical orbits). A one-way trip by rocket from Earth to Mars would take a curving path using the gravity of the sun during its middle segment. It would take at least nine months, and longer when the distances are greater. (The longer times are relevant in terms of rescue when something goes disastrously wrong.) Mars has long seasons, and the temperature varies greatly between them. It is always much colder than Earth, and in winter can get as cold as −87 °C. It has extremely little oxygen in its atmosphere, which is mostly carbon dioxide. It has no standing water. It probably has some amount of water in frozen form beneath the surface, and perhaps a fair amount, but the amount is not known. We have not yet found any form of life on Mars, though it is far from inconceivable that it once existed. Mars has no overall magnetic field, which has the consequence that ultraviolet radiation from the sun and cosmic radiation from all sources are stronger than on Earth. In particular, there is a greater level of radiation in forms that would be more dangerous to human life than there is on the surface of Earth. Solar energy, which would provide the main source of power for a colony, is considerably weaker than on Earth and is interrupted by dust storms.

The book and the film *The Martian* depict a forceful, well-educated astronaut surviving on his own when he is abandoned on Mars. It is well plotted, and the hero is a tribute to human ingenuity. But a number of things are left out. We are not shown the low light level; the ferocity of the dust storms, especially at night, is minimized; there is no suggestion of the extreme cold and the dangers of radiation and lesser gravity – not to mention the hazards of living on potatoes for 18 months.

All this makes Mars seem a very unpromising location for human habitation. And there is no denying the enormous challenges of survival there for members of our species. So why choose Mars as a refuge? The answer is that other locations in the solar system are even less promising. Less promising for human habitation, that is: not necessarily less promising for life in general, especially if we draw the boundaries of what counts as life quite widely. Planets nearer to the sun than

Earth are impossibly hot for humans; planets further away are impossibly cold. The moon (Earth's moon) has enormous temperature variations, no atmosphere at all to inhibit meteor and asteroid strikes, and long periods of darkness except at the poles. (The moon has possibilities as a staging post for further exploration and as a location for scientific bases. I would think that the moon is the most promising place in the solar system for human habitation after Mars, and it is much nearer – after Earth, of course.)

The plans

No organization of any government has plans for Mars colonization. Bodies such as NASA have limited and uncertain budgets which are fully committed for scientific and other purposes. Both the United States and Russia have somewhat vague plans for human missions to Mars in the 2030s, but colonization plays no part. NASA would prefer to use its limited budget on satellite and robotic exploration and on planetary protection. (There are plans for one more NASA rover, and there are Chinese, European, and Canadian plans.) They may also have calculated that the risks are too great, especially given the harm that a catastrophe involving loss of life would have for their public status. The initiative for human Mars travel has passed into private hands.

The most prominent initiatives are the Dutch Mars One and Elon Musk's SpaceX (Mars One plans to use spacecraft from SpaceX). Mars One has stated an aim to send unmanned missions in 2018, cargo in 2022, humans in 2025, and four more people every two years, with the intention of having twenty colonists by 2040. Volunteer colonists have already submitted their names.

Many people were initially enthusiastic about the Mars One plan, attracted by its promise of relatively modest costs and independence of public funding. Closer examination of both the proposed technology and the budget has disillusioned many. A few have even suspected fraud. A typical reaction, but probably better informed than most, is that of Gerard 't Hooft, who shared the 1999 Nobel Prize in Physics. He was initially a supporter of Mars One but more recently has been quoted as saying that a launch date 100 years from now with a budget of tens of billions of dollars would be an achievable goal. The combination of considerations about cost and considerations about timing is significant: haste makes waste.

Elon Musk is well known for developing advanced technology that addresses social, especially environmental, problems and also makes a profit. The profits are partly intended to fund the research behind the Mars project. His company SpaceX has been developing rockets and space vehicles for use in private spaceflight. It has undertaken a supply mission to the international space station. One feature of the company's rockets is that they are meant to be reusable. Its smaller rockets have been reused. Of course, repeat trips to Mars with giant rockets are completely untested.

At the moment, SpaceX's plans for sending people to Mars are not very detailed, and there is no timetable. As of the time of writing, fall 2017, Musk is talking in terms of 2024 for the earliest possible date. Of course this may be postponed. Although Mars colonization is among Musk's central ambitions, the development of reusable rockets and spacecraft is a more immediate priority. Future projects that have been mentioned include experimental vegetable growing on Mars and, more recently, experiments to manufacture rocket fuel there. (The advantage of making fuel on Mars is obvious: unless it is possible each voyage must have enough fuel for the round trip.) These experiments would be carried out by robots, with no risk to human life, and so the attempts could be abandoned if they were failing.

Compared to those of Mars One, SpaceX's plans are modest and cautious. Yet there is an element of extreme ambition about them: the plans may not be feasible, or not within anything like the envisaged time-frame, but if they were to succeed something dramatic would have been accomplished, especially given the large number of colonists that are part of the plan. The caution makes sense: no firm deadline for the actual colony is given, and each preliminary stage has to be passed before the next is tackled.

The aspects of SpaceX's plans that are undeveloped, not revealed, or simply hopeful make it hard to give a detailed evaluation. My discussion of SpaceX continues in chapter 4, where I try to tease out some probable costs. For now, I shall just list the most important things that do not seem to have detailed plans yet. The most important is the habitat, where people are to live and work and produce food, as discussed in the next section. Then there is food production, raising plants in soil based on what is available on Mars, or in other ways, and producing food in artificial cell cultures. As far as I know, plans for this and the choice of corresponding technology are completely undeveloped, though the topic will have impacts on provision of indoor space, heat, light, and oxygen. Power production will be based on solar energy, though details of the production and design of solar cells, and how they will function with lower light levels and frequent dust storms, have not been specified. It is worth noting that power is needed not just for industrial activities but also for the production of heat and oxygen. Industrial activities include the manufacture of fuel from Martian soil, something whose details may have to be left to future robotic experiment. The last vital element is oxygen, which is very scarce on Mars. It will also have to be produced from the Martian soil, which is certainly feasible but energy intensive and potentially dangerous, so the details of the plan are important.

Given the lack of detail, the situation is difficult for someone trying to produce an informed commentary, as also for anyone engaged in a public discussion of plans that could risk numbers of lives and appropriate significant resources. On the other hand, it would be irresponsible of those at SpaceX to announce plans based on untested technology or to commit themselves to solutions that may well be changed with more experience. So there are pressures both ways. The obvious conclusion is to withhold judgement until we know a lot more.

Habitats

No colonization will happen until there are workable plans to shelter the colonists. A habitat for the settlers must provide oxygen, heat, food, and drinkable water. None of these will be available locally and all of them will need to be produced for years. A major complication is the lesser atmospheric pressure on Mars. The interior of any habitat will have to be Earth-like and so at a greater pressure than the exterior, leading to a danger of explosion if the structure is not solid enough. But the easy way to solidity is to use heavy materials, which would be prohibitively difficult and expensive to transport from Earth. So construction from local materials is usually suggested. This construction would almost inevitably be done by robotic machinery sent in advance. The detailed work would draw on 3D printing from prior plans. But saying this leaves the design open. What might a workable habitat be like?

NASA has been running a competition for designs for a Mars habitat that can be constructed in advance. The competition has reached phase 3. Phase 1, whose winners were announced in 2016, was for general designs, amounting really to architectural drawings. Phase 2 was at the other extreme – plans for 3D printing of structural elements from recycled materials and material available on Mars. Winners have been announced. Phase 3 will concern construction plans.

The most significant feature of this competition, for our purposes, is how far from realization the plans are.

We are still trying to figure out good ways of producing the components of habitats from local material. A colony continually replenished from Earth could import lighter or more sophisticated components, but a self-sufficient colony would have to produce them all itself. So, until this is a possibility, the colony-as-refuge idea will be extremely incomplete. Assuming that NASA and the organizations with which they are cooperating have the most advanced and practical ideas about habitats, this fundamental aspect is still at a relatively early stage.

One alternative to freestanding habitats that is sometimes mentioned is to use natural caves or lava tubes near the surface of Mars. This seems to be in the category of interesting idea rather than worked-out plan. Presumably there is still a need for construction with local materials, but some problems, such as radiation shielding and the issue of the inside/outside pressure difference, would be reduced.

Some of the techniques that would be needed to construct habitats on Mars would also be required for other purposes. Equipment will always need repairing and eventually new equipment will have to be produced. If a large number of spaceships transport colonists and make the delicate landing, which has taken a lot of ingenuity with the much lighter rovers, then it is likely that some of them will be damaged and need repair. Some industrial capacity may be necessary from the beginning.

Terraforming Mars

Terraforming is the process of changing conditions on a planet so that it is more suited for life that has evolved on Earth. In the case of Mars, this means changing the atmosphere so that there is more oxygen, increasing the temperature, and creating a magnetic field to reduce radiation and help retain the atmosphere.

The first two of these tend against one another. The easiest way to raise the temperature is to produce even more carbon dioxide, leading to a greenhouse effect of the kind we are trying to avoid on Earth. But the proportion of carbon dioxide would not be decreased by more oxygen. A complication is that Mars is smaller than the earth so that its gravity is less effective in retaining an atmosphere. Other greenhouse gases besides carbon dioxide have been suggested, but the possibilities are not well understood. In any case, the level of oxygen that could be achieved using present technology over a long span of time would make the atmosphere some-what similar to the high Himalayas, where people can survive but need extra oxygen.

The Martian soil is an obvious focus for terraforming. Mars does not have the history over millions of years of absorbing organic materials, which are then processed by earthworms and microbes and other living creatures, to produce soil as we know it that makes agriculture possible on Earth. It would be like farming the desert. There would be no point trying to rectify this except in small controlled habitats until the planet was warm enough and its atmosphere had enough oxygen to sustain these in an Earth-like way. But until something like this is achieved, all agriculture will be in miniature. Work on artificial fertilizers and artificial soils – for example analogues of the hydroponics that are sometimes used for urban farming – is needed.

There are various ways of creating a magnetic field, and there is no consensus about the best way of doing it.

Estimates of the time it would take to achieve even modest aims, which still leave Mars a very different place from Earth, vary from slightly less than 1,000 years to 100,000 years. The core reason why the estimates vary so much is that we do not understand these things well. Climate modelling for the planet where we have the most data, Earth, is still in a relatively early stage and its predictions are fairly imprecise. Modelling enormous changes by untested means for a less familiar planet is inevitably even more uncertain. In any case, given that even 1,000 years is longer than most human plans, there is time to learn more and get it right – if this is what we want to do, that is.

These are long-term plans. But it is remarkable how they throw light on factors that would affect the survival of a colony from the very beginning. Breathing, eating, and moving about are hardly optional aspects of human life. All these problems can be solved. But they need to be tackled in the right order and on a suitable timescale.

Terraforming raises issues of contamination of Mars with Earth life in an acute way. Human colonization of Mars would inevitably lead to some contamination, and perhaps a lot of it. Remember that we are swarming with one-celled organisms on our skins and in our guts, and that microorganisms are better than more complex life at adapting to hostile conditions. But a terraformed planet would mean that much less adaptation would be necessary. This would probably be the end for any native life on Mars. (Prospects for its survival are pretty dim in any colonization scenario anyway: humans now have a chance to extend their species extinction activities to other planets.) The only safe conclusion is that any interaction between humans and the planet

Mars will require extreme measures to preserve any local forms of life. Perhaps Martian life will need a special refuge of its own.

Legal issues

Whether colonization of other planets occurs as a result of private initiative or government action, difficult legal issues arise. Who owns the land, and what does it take to establish ownership? What limits are there to the exploitation of resources? Who makes the laws? Could one Earth country claim a whole planet? European colonies in the Americas and Australasia acted as if the people already there had few rights to the land and its contents. It is generally recognized now that this was atrocious. There is an interesting contrast here between the views of the astronomer Neil deGrasse Tyson, who argues that private initiative rarely acts out of adventurous idealism, and those of Elon Musk, who argues that government organizations are rarely innovative. Both could be right.

The UN space convention goes some way to providing what is needed here. It demilitarizes space, subjects private organizations to the control of their national governments, and prevents any state from claiming a celestial body. Many, but not all, national governments have ratified the convention. It does not apply to the moon, and a separate treaty covering the moon has not been signed by any national government with the capacity to violate it. This does not augur well for the prospects of the existing convention holding when national rivalry or commercial interests become pressing. There is no provision for enforcement.

Some believe that the convention is too weak, allowing an inadequate basis for exploitation of extraterrestrial resources. Others believe that it is too strong and inhibits legitimate commercial activities in space. It will be a while before these issues become acute, but there is a lot to be said for settling them before they become really charged.

There is another very important legal issue, which is the rights and governance of colonists. Are they to be considered as citizens of a little republic on Mars? If so, can they change the decisions of the organization that sent them there? Or are they subjects of that organization and subject to its rules whether they like them or not? They will be told that they are embarking on a one-way trip and will spend the rest of their lives on Mars. But can they be held to this? Can they rescind subservience to the organization? Many contracts on Earth are not enforceable because they violate basic rights. (I could sign a piece of paper making myself someone's slave, but that person could not legally compel me to abide by it.) So would the same be true of colonists, or would it be more like joining the army?

The paragraph above consists almost entirely of questions. There may be better and worse answers to them, but even bad answers are better in this case than no answers at all. For, given definite answers, people will know what they are signing up for. They may then regret the decisions that they make, but they will have entered into them with their eyes open.

Timescale

Technology needs to be developed, decisions have to be made about the protection of Martian life, and clarity is needed about the rule of law. If colonization is to go ahead soon, then all this must happen in a hurry. I doubt that it can happen fast, so my hope is that delays in the development of the technology will give us time to think out some of the other issues. There is a lot to be gained by taking time to get these things right; basic decisions are likely to look very different if they are thought out slowly and carefully. In any case, issues of timescale, and of the order in which problems are tackled, are crucial. It is one of the things that we should think very hard about before doing anything irreversible.

THREE

Problems with Colonies

All of the solutions that have been proposed for threats to the human species have problems. No surprise there: no human solution to a human problem is perfect. We have to state these problems as clearly as we can and then estimate what it would take to resolve them. Stating the problems is the purpose of this chapter, and estimating costs is the purpose of the following chapter.

Worries about refuges

To be refuges where humans can survive catastrophe on Earth, colonies on other planets must of course contain and sustain humans. That is the point. They must also be highly technological: surviving in an environment less hospitable than anywhere on Earth would need powerful resources. Mars does not have an atmosphere that we can breathe, does not support plants that we can eat, is very cold, has little usable water, and receives much less solar energy. It is hard to make an analogy with anywhere on Earth: combine the light levels of the deep ocean with the cold of the Antarctic, add radiation, and then exaggerate. (The pictures from the Martian Rovers are accurate as far as colour and illumination go, but we tend to project familiarity onto them, taking the atmosphere to be like air on Earth and reading the absence of snow and ice as warmth rather than the frozen desert that it really is. I know this is my own tendency until I catch myself.)

The colony must from early on produce all its own food, water, and oxygen. This is not at all impossible, given sophisticated equipment, which has been tried out under desert and arctic conditions on Earth. But these conditions are not really that much like Mars, especially with respect to cold, dark, and radiation. The equipment must continue to function, indefinitely. So it must be possible to repair it without using supplies brought from Earth. So, until local manufacturing can take over, repair equipment and spare parts must be added to the list of things that must be sent with the colonists in the first place. And, easy to overlook, it adds to the number of people who must be sent. A modern technological society of a kind that can create and repair the kind of equipment we are talking about involves thousands of specialized skills. Some combinations of these can be compressed into a smaller number of people, but many are still needed. Robinson Crusoe would not last long on Mars.

Questions about the number of people in a colony are crucial. Self-sufficiency requires a large number of people – say several hundred at the least. And long-term survival requires genetic diversity. If population sizes are too small, then inbreeding makes hereditary defects and infectious diseases more common. Moreover, with a small population size, random fluctuations can result in imbalanced numbers of males and females, leading to both a smaller number in the following generation and yet more reduced diversity. (A shortage of females is obviously more serious. A bias towards females would have obvious advantages. Perhaps in fact an ideal colony should be all female plus a genetically diverse sperm bank.) It has been estimated that in wild quadrupeds a population size of 500 to 1,000 is needed for long-term survival of a species, while the crews for the simulated Mars habitats on Earth have typically had six people! Humans already have a very low genetic diversity: pairs of chimpanzees in the same troops have on average more genetic diversity than pairs of humans on Earth.

The crews would have to be carefully chosen. A very special psychological makeup is needed. Crew members must endure close quarters with a small number of others, a very basic life, the knowledge that one has left one's family and friends behind, and a high risk of death. They must also be chosen so that there is a range of technical knowledge, improvisational skills, and the emotional and cultural makeup needed for something like Earth civilization to continue. And this must reproduce itself for generations. It is unlikely that, even if an optimum mix of people were achieved in the initial crew, the same mix would be preserved in subsequent generations. This too argues for larger population sizes. But the more people there are, the greater the expense and resources needed to establish the colony in the first place.

A disturbing fact about the production of food on Mars has recently emerged. The soil on Mars is rich in compounds called perchlorates. They react with ultraviolet light, to which the Martian atmosphere is largely transparent, in a way that is fatal to many cells. There is thus a lot of doubt whether plant crops, and the symbiotic bacteria that many of them need, can survive in Martian soil. This complicates ambitions for indoor farming considerably. Because of the effects on both living cells and human health, perchlorate contamination is regarded as pollution on Earth. Perchlorates also have a risk of explosion when they are heated, complicating plans to produce oxygen by heating the Martian soil. They are, however, a source of oxygen and of other basic chemicals; although dangerous they could have their uses. There are surely high-tech solutions to this problem, but equally surely they raise the stakes for transport and technology and increase the danger.

The complexity of technological society

There is a fundamental fact behind many of these problems: the large scale and interdependence of our society, with its complex web of manufacturing techniques and expertise held in the minds of many people. It is extremely hard to duplicate this in a small population with restricted resources, especially in a hostile and unfamiliar environment. So dependence on the mother culture is hard to avoid. (This was true in the past, also. The early European colonies in North America did not make their own muskets until they had grown quite large, and European agricultural styles took a lot of adapting. This may not seem advanced technology. But could you make a musket? For that matter, could you make a stone axe?) This means that the high-tech devices needed to survive in the Martian environment are not going to be designed there. The designs are going to come from home. And it is likely that at least a proportion of the devices themselves will also.

44

3D printing from transmitted designs may solve some problems, though, *if* the raw materials can be obtained and refined on Mars. (I would imagine that supplies of direct and indirect biological material, such as the petroleum and oil products that are used to make plastics, might pose a serious problem.) If imported equipment is unsuitable or does not work because of some unexpected quirk of the faraway environment, much of it will have to be redesigned and manufactured not where it is needed but where the techniques and expertise are to be found. The more advanced the apparatus (the higher the tech), the more will need to be transported to the colony, adding to the transport costs and creating a need for spares.

For all these reasons I am extremely sceptical that a colony of the size that we could send to Mars in the next decades, perhaps in the next century, could sustain itself without frequent supplies and reinforcements from Earth. The obvious reply to this is to drop the requirement that the colony be able to survive without the supplies and reinforcements. But this would undercut one of the main purposes – that of providing a remnant of humanity on Mars with a reasonable chance of surviving an earthly catastrophe. The colony would then be a scientific expedition and the beginning of a preparatory project that might take centuries.

Three kinds of catastrophe

Many things can go wrong for the people occupying an established base on a planet, whether a colony or a research station. There are physical catastrophes, medical catastrophes, and social catastrophes.

Physical

Asteroids can strike; holes can develop in super-secure walls; equipment can fail for no known reason or because of errors in manufacturing; unprecedented bursts of radiation can occur. If I could give a complete list then we would know what to anticipate, but we must assume that unanticipated things will happen. Some of these – breakdowns in physical barriers, problems about oxygen or food supplies – can produce widespread or universal death. Others can produce miserable or desperate conditions until a rescue can be organized.

These problems can compound one another. A freak accident can cause the death of the only technician able to repair a vital piece of equipment. An asteroid can obliterate the area used to store food or crucial catalysts. It may not be possible to anticipate all of these, and forestalling all of the improbable ones would mean a large population and a mountain of supplies. But it takes a smaller catastrophe to wipe out a colony than a planet-wide species.

Medical

A virulent epidemic can carry off a large proportion of the population. Perhaps it is due to a maliciously transported virus, perhaps to an infection with a very long incubation period, perhaps to a mutation in normal human symbiotic bacteria that would not have occurred on Earth. In the limited space, placing affected persons in quarantine may not be possible, or the outbreak may be detected too late for quarantine, or the outbreak might be in a particularly essential work area, and the medicines for treating it have not been included in the supplies.

Cancer rates may be higher because of increased radiation. They probably would be. If by chance they affected essential personnel there would be trouble for everyone. Evidence is accumulating that radiation may actually be the great limiting challenge. The standard human microbiome, the vast numbers of one-celled organisms that live in our guts and on our skins, presents another unknown. It might vary or fail when the radiation and the gases we breathe are different and when we are not surrounded by animals, plants, and microorganisms with their own symbiotic colonies, leading to metabolic, immunological, and other problems. Suppose that, as a result, a large proportion of the population develops a chronic disease: the consequences would be unsustainable.

Radiation levels, from the voyage and in the colony, may induce sterility. It may then become compulsory for the remaining fertile people to have numbers of children if the colony is to maintain its minimal viable number – another restriction on individual freedom and satisfaction (*The Marsmaid's Tale?*).

Most of these medical troubles can occur on Earth also. But they are compounded in a station far away. The population is smaller and does not have the variety and redundancy of skills and genetic constitutions. Advanced medical techniques and sophisticated drugs will not be available until there is a large well-established colony. The space is more limited because most activities are confined to a constructed settlement. (Think of a virulent and easily spread virus on a cruise ship.) Notice how the interlocking of many skills and forms of specialist knowledge in modern society is relevant. Notice also how medical and ecological facts that we do not understand about life on Earth are also relevant. The two are connected, since among the things we do not understand about human life is how it can be downsized to smaller populations. And intricate ecosystems are extremely hard to understand in detail.

Social

Any group of people needs direction and decision-making. Some of the people in such a station would have come from a military background and others would be used to a more fluid style. These differences could lead to very serious trouble when combined with physical or medical catastrophe, necessitating drastic decisions such as allowing stricken people to die or abandoning a central aim of the expedition. These may be bitterly or even violently contested, and the result may be a breakdown in social decision-making. They also portend trouble in merely miserable situations. At first people will say "We are on Mars! Isn't that wonderful?" Then it will slowly sink in that they are living in cramped quarters which include others they would like to get away from, eating a very basic diet and having to provide most of their own entertainment, following orders about what they must do and unable to choose freely who they love, whether they have children, and so on, unable to explore the planet at all freely – in fact, very rarely going outside – and with a constant threat of sudden death. Some will react with depression and others with anger. And those born in a colony will not have been selected for optimism and resilience.

These unhappy people will try to find unauthorized ways of returning to Earth. This may be violent; security forces may be needed. Some will send messages to Earth saying "Don't come here." These attempts may provoke fierce responses, and catastrophic damage may occur. An image of contention alone will inhibit future immigration, which will reduce genetic diversity. (This would also be endangered by radiation-induced sterility.)

Largely indoor life in cramped quarters, eating unvaried vegetable and cultured food, constrained to follow occupations that are essential for the survival of the colony, little individual freedom and no accumulation of wealth: it is hard to know who would be able to adapt to this, even among carefully selected volunteers.

Children would be born in the colony. Reproduction presumably would be forbidden in a non-permanent expedition and would have to be thought about carefully in a colony sustained from Earth. But for a colony that sustained itself in the long run, new generations are obviously essential. And there is the need for genetic diversity. So babies would be encouraged, and schools and teachers would be needed. People would also age and need to be supported in their old age. There would be illness, so hospitals and doctors would be required. These might be even more necessary given the greater risk of radiation-induced cancer from spaceflight, the greater radiation on Mars without a protective magnetic field, and the risks of muscle wastage and osteoporosis from the weaker gravity. This would affect both exploratory and refuge colonies; some colonists would arrive in very bad shape. In a closed environment contagious disease is a great danger, so an isolation area might well be needed.

Because children born in the colony would be as varied as human beings are, and because of the novel social stresses, educational, psychiatric, and criminal institutions would be necessary. Order would have to be maintained, especially if the stresses of life in an environment humans have not evolved for proved to be great. How coercive this might be is hard to assess, but well-intentioned anarchy would surely not be enough. More people, more areas, more things that can go wrong.

"But we have tested Mars habitats on Earth, in the desert and elsewhere." Not really. Teams that are much smaller than the size of any colony would have to have spent periods of a few months in simulated Mars structures. These do not test the effects of low gravity, lesser sunlight, extreme cold, radiation, and dust storms.

And, if the worst comes to the worst, those taking part in simulations can just step outside, as they did on one occasion when the habitat burned down. The primary aim of these experiments is to test for social problems, to evaluate the possibility of doing scientific research, and to produce food. And in these terms they have proved satisfactory. However, the production of electricity from Martian sunlight levels and its use to manage oxygen, water, and heat in a Mars-like environment has hardly been tackled by these simulations. The radiation problem has not been tackled at all, and it is hard to see how it could be on Earth over a short span of time.

Is there safety in distance?

Suppose Earth is convulsed by war or plague. Would a colony on Mars really escape? Wars are usually driven by human irrationality, and the desire to punish the opponents dominates. Why not target all members of the hated nation, religion, or ethnic group whatever planet they are on? An asteroid-deflecting nuclear missile could be nicely adapted, as would an engineered virus. This would be insane, but human insanity is just the danger in question. An international trans-human colony would escape some hatreds and provoke others. Certainly a nationality or ethnic group that seemed to evade the fate of all other humans would risk provoking dangerous if irrational resentment. A colony would depend deeply on its computational capacities, and these would be vulnerable to malware like that which the United States, Russia, North Korea, and Iran successfully direct at one another, in spite of sophisticated defences. A Mars colony could be wiped out by a simple computer failure.

Computer failures are a fact of life, anyway, and any colony will have taken measures to minimize their likelihood and their effects. There is a connection with one of the motives that is sometimes given for establishing colonies on other planets: the danger that rogue AI will turn on humanity. I argued in chapter 1 that this danger is exaggerated. But in any case there is something very strange about going to another planet to escape artificial intelligence. On Earth there is the possibility of disconnecting or sabotaging threatening technology. People can live in the woods for weeks. Mars would be different; humans could not survive without their high-tech tools. And many of the motives for cybernetic takeover would apply to distant humans also; it is surely not beyond the capacities of a machine that can organize a cybernetic putsch to extend its ambitions beyond the planet. Telecommunications would be the obvious means. And, at least in the initial stages, computers in the colony would be linked to computers on Earth, for example to download and present solutions to complex problems and for 3D printing. The colony would hardly be a refuge from any sinister designs of earthly computers. On the contrary, it would be at least as vulnerable as human civilization on Earth.

Issues about the future capacities and dangers of AI and other forms of robotics permeate questions about colonization and space exploration. They crop up almost everywhere. It is obvious that our ambitions to colonize and explore put intense demands on powerful and trustworthy computation. And we just do not know what demands are going to be met on what timescale. A central issue that crops up throughout my discussion is how many human intelligent tasks can be performed, and performed better, robotically. There is a paradox here: the more capable computation becomes, the less need there is for human travellers; but the more we humans want to travel, the greater our need for powerful computation.

An unknown here is the possible climate and environmental damage from all those rockets. We do not know enough. In the worst case, establishing a colony would aggravate conditions on Earth that encourage war, famine, and disease. I discuss this in chapter 4.

The "risk compensation" effect, where safety measures lead people to take greater risks – for example, driving faster when they have a seat belt – also applies. A national leader considering whether to retaliate for a non-devastating nuclear first strike who thinks that, even if the worst comes to the worst, there will still be humans surviving beyond the planet may be more likely to push the button. (This is akin to what economists call "moral hazard": insurance makes it less unreasonable to take risks.) Nuclear arms have not been used since 1945. One reason that is often cited is the probability of destruction of both sides and more. But this is undermined by even slim chances of further survival, in particular survival of the national group which would otherwise be deterred. In the very worst case, human life on Earth is ended and then the colony on Mars, or wherever, also ends. So its existence will have been one of the causes of the end of the species. As the astronomer Lucianne Walkowicz has said, "it is like the captain of the Titanic telling you that the real party is on the lifeboats."

The point is very general. The greatest threat to human life, in the large and in the small, is other humans. Sending humans elsewhere does not change this. The problems of hatred and craziness have to be solved, wherever we choose to solve them.

Resources

Perhaps all of these problems can be solved with enough time and money – with enough of the world's resources. But they suggest that the problems are deep-seated and not easily solved with present technology. We have to think carefully about costs, and especially about opportunity costs: what other aims would be hindered by solving these problems. And we have to think carefully about timescale, about how quickly we want solutions and the order in which we want them.

FOUR

Costs of Colonization

Everything has a price tag, and any price is worth paying if the benefits are great enough. The tangible benefit suggested for colonizing other planets is escape from the possible fate of humanity on Earth. Less tangible benefits are scientific knowledge and human exploratory adventure: the pride of our species. But these are definitely not the only ways these benefits can be obtained – or even the best ways of getting them. Costs can be measured in money, which has immediate effects on standards of living. But we must not ignore opportunity costs – projects which are likely to be neglected if our resources are directed in particular ways. Begin with money.

Money: Mars One

I shall concentrate on Mars exploration. NASA has successfully landed three rovers on the planet (before that there was one NASA failure, two Russian failures,

and a European failure). The latest and most successful rover, Curiosity, weighing 900 kg, has cost some
US$2.5 billion. For comparison, the international
space station (ISS) had cost $150 billion up to the year
2010, perhaps the most expensive object ever made.
It is vastly larger, weighing 420,000 kg. So, although
all these figures are very rough, the rover is something
like 8.4 times as expensive per kilogram as the space
station, even though it has no human crew. At this
rate, constructing and shipping to Mars something the
size of the ISS would cost more than $1 trillion ($10^{12}$:
1 followed by 12 zeroes). But the ISS has a maximum
crew of six, so scaling it up to 1,000 people to give the
range of skills and expertise and the genetic diversity
needed for a permanent colony would give a figure
of more than $167 trillion. Though a colony on a
planet and an orbiting space station are very different
things, the weight and cost comparison could go either
way. The ISS ships all its food and replacements for its
machinery from Earth, which would not be possible
for any permanent establishment on Mars. A colony
would need farming or other food production areas
and manufacturing and storage areas. So a Mars station
might have to be three or four times as big per person,
which would take the cost to near $1 quadrillion ($10^{15}$).
The US defence budget in 2017, for comparison, was

more than $500 billion. (The figure is out of date, and there are costs not included in the official military budget. What proportion of the American GDP goes to defence is very hard to ascertain.) Several quadrillions is not unmanageable if it provides the only solution to a life or death problem. But it would be a severe economic burden, and I would expect it to be politically impossible.

One of the two private initiatives, Mars One, has cited a cost of US$6 billion to achieve the aim of four colonists in 2031 and twenty by 2040. This cost is widely regarded as unrealistic. One absolutely essential fact here is that these are planned as one-way trips. No return. Since the expedition is one way, and since technology and techniques for supplying oxygen, food, water, and maintenance to a Mars colony are even less developed than ways of getting there, this will involve many continuing expeditions to the planet. An independent MIT assessment of the costs of the Mars One plan concludes that "the amount of mass that must be sent to Mars at each resupply opportunity increases with the number of crews on the surface." The reason is that the first crew "must bring enough spares to sustain themselves for 26 months. The second crew must bring enough for themselves and the first crew, the third crew must bring enough for themselves, the second crew, and the first crew, and so on." The same analysis continues: "By the tenth mission, the launch cost in the BPS [indoor farming] case is approximately $12.6 billion, and the cumulative cost of launches to grow and sustain the colony is approximately $109.5 billion. For the SF [stored food] case, the cost of the tenth mission is approximately $15.6 billion, and the cumulative launch cost is approximately $106.8 billion."

A basic reason why resupply is essential to this and other colonization plans is that, given the cold and the lack of oxygen, human survival on Mars is a high-tech business. Space One's plans for producing a breath-able atmosphere in the inflatable habitats of the colony involve heating Martian soil, which, it is hoped, would produce enough water to be decomposed by an elec-trolytic process to produce oxygen. This, together with producing food in a confined environment, recycling wastes, and conserving water, requires sophisticated machinery and a range of expertise. When things break they must be replaced, and if people die or are ill someone must have their knowledge and skills. Sending replacement equipment to Mars would take from nine to eighteen months, depending on the orbits at the time, and colonists might have starved, frozen, or suf-focated by then.

The Mars One plans for food, breathable atmosphere, and energy have been particularly criticized. Self-sufficiency in food would require vegetable-growing areas, which of course also would have heat and atmosphere. But plants produce oxygen, and without much nitrogen in the atmosphere the danger is that enough plants to feed the colonists would lead to a fire hazard from the oxygen concentration, especially given the perchlorates mentioned in the previous chapter. Plants also need light, so sufficiently powerful electric lights would be needed in growing areas, powered by solar cells that would also power the electrolysis of water to make oxygen. Sunlight is less intense on Mars and there are long-lasting dust storms. Thus very efficient solar power collection and storage would be needed. The dilemma with advanced technology is that each possibility comes with a problem: shipping equipment has delays and adds to costs, and making something on site requires that there already be an industrial base and a range of expertise.

Two central lines of criticism of the costing behind the Mars One plans emerge from this and other discussions. Both stem from the same general problem with planetary colonization. One criticism is that the plans make unrealistic assumptions about how easy it will be to solve various technological problems to do with the production of food, water management, and the provision of a safe breathable atmosphere. The other is that the total mass of supplies and equipment that would be needed is much greater than estimated. Together, these suggest that the initial and continuing cost is much greater and that the costs and the probability of disaster are increased by haste. There are fundamental problems that have to be solved before this is anywhere near workable.

Money: SpaceX

A central aim of Elon Musk's company SpaceX is to bring down the costs of establishing a Mars colony. It is hard to get firm figures about this from any source. Current NASA techniques pushed just a little suggest a price tag of $10 billion per colonist, which is clearly an enormous amount for the number of people needed to have a viable colony. Musk's ideas for making the project financially more manageable, especially if it is not publicly funded, involve reusable vehicles that are refuelled in Earth orbit and on Mars, obtaining fuel on Mars, and, most dramatically, sending a large number of colonists using reusable vehicles which make round trips and do not have to be replaced. This does reduce the resource requirements for each vehicle. But there would be a lot of them! It also makes the use of fuel obtained on Mars essential, and we do not yet know how to do this. At the time of writing, Musk proposes that the same equipment can be used for intercontinental rocket travel, thus subsidizing the Mars project. (A plan to increase the capacity to 200 passengers – still 5,000 trips – seems to have been dropped.) He has stated that, if a million colonists are transported, the price can come down to $200,000 each. The vehicles will take "just" 100 passengers, though, so this means 10,000 flights, whose environmental effects obviously have to be investigated. No spacecraft of anything like this capacity have ever been constructed, or even designed. The rockets would be larger than any yet made and would refuel in Earth orbit. The calculations behind these figures have not been made public, and we can imagine that they will change as plans progress.

SpaceX's plans are cautious compared with those of some of its rivals, and the costs are presented as tentative. Still, it is clear that a lot of money is involved. It is hard to put a total cost on plans like this, especially since, given the resupply problem, there are continuing costs. A million people at $200,000 each is $200 billion, and we can expect resupply costs for years after settlement. It is unlikely the costs can be brought down by reducing the numbers, since the low $200,000 figure was obtained as a specific economy of scale. Let us assume that the minimum size of a colony is 1,000 people. (I suspect this is an underestimate. For comparison, the British colony at Jamestown was not viable until it had several hundred people, in a much more hospitable environment where people already lived and flourished, and with much less reliance on technology.) At the old $10 billion per colonist figure, this would come to $10 trillion – rather higher than the Musk figure for a million colonists.

Besides the seriously untested technology, there are factors here that are very hard to balance. The bigger the colony, the greater the resupply problem for items and material that cannot be produced on site. But the bigger the colony, the greater the available range of skills needed to produce and repair things on site. This is in part a sociological problem: what is the irreducible range of skills and specialist education needed to maintain a technologically advanced society? We don't know the answer to this, which is at least as hard as most scientific problems. (Never trust a quantum physicist to fix your television. And banish the thought that all hard problems can be given physics or engineering solutions.) So the numbers and resupply question is extremely uncertain. There is the possibility that even a million-person colony would not be self-sufficient and would suffer a catastrophic failure which could only be rescued at astronomical expense. After all, Singapore, a technologically advanced place, has a population of 5.6 million and is not completely self-sufficient.

What in the end are we to say about costs? I have mentioned figures from $200 billion to $167 trillion and suggested circumstances that would raise the upper figure. These are in line with the MIT study of these issues. They are not unmanageable amounts for current advanced economies (though the costs of recovering from a major miscalculation would be another thing). Of course there are also continuing costs. No one is thinking seriously of any date before the 2030s for such expeditions, so by then things might be more focused.

This is an argument for preliminary studies, experiments, and pre-colonizing expeditions. Such preparations are also required to minimize the probable human cost: deaths, misery, people stranded in awful conditions for long times. We will obviously pay to avoid these, and if they threaten we will be compelled to pay what it takes to avoid or recover from them. This is especially true for colonists rather than professional astronauts, who willingly, in fact eagerly, go into a dangerous profession. So human costs and financial costs are intertwined.

Astronauts are likely made of sterner stuff than most colonists. But colonists will not be naïve and will have some sense of what they are getting into. And on Musk's plan they will be paying their own way. (The prospect of surviving the end of human life on Earth is dangled before a few well-off adventurous spirits – not typical humans.) So, making some enormously risky assumptions to give such plans the benefit of the doubt, assume that there are enough potential volunteers in the face of facts such as those I have listed in the previous chapter, eager for a one-way trip to live on basic rations in cramped quarters for the rest of their lives, with little guarantee of rescue if things go wrong. Assume that no continuing subsidy is needed, and that there is no burden on the economy of Earth – for example, from lost taxes. If we dare to assume all this, the burden passes to environmental and opportunity costs. If the colonists are supplying their own money and lives, what does it cost the rest of us?

Environmental costs

What would it do to the earth to send all these expeditions to Mars? There has not been a great deal of analysis of this. We can divide the environmental costs into two parts. There is the environmental cost of producing the spacecraft, the rockets that launch them, and the equipment they carry. And there are the environmental effects of launching them.

A reasonable supposition is that environmental costs of manufacture are proportional to financial costs. Industry is industry. Suppose that a Mars colony costs between US$5 trillion and $50 trillion to establish and then between $10 billion and $100 billion each year to maintain until it becomes self-sustaining. This is a stab in the dark; no one knows. We could make discoveries that bring the price down, and we could also find unexpected hazards that raise the cost enormously. For rough comparison, the sum of the gross national products of the countries in the world in 2014 was estimated at $78 trillion. As mentioned above, the US military budget in 2017 was $596 billion (or more, depending on what you include). A single aircraft carrier, the USS *Ford*, is expected to cost $13 billion. The US gross national product in 2015 was $18.14 trillion. So a 10 to 100 times increase in the US defence budget, in terms of tax burdens or contributions from individuals, or between 6 and 60 percent of the production of the earth, could pay for it. So we are reckoning on an environmental burden proportional to this. Roughly!

The major costs would take the form of industrial production, often heavy industry. Since industrial production is much more damaging to the environment than the service economy, the environmental burden might well be greater than the proportion of GDP would suggest. And further reductions in heavy industry are frequently thought to be necessary for the health of the planet, since the present level is not sustainable.

Just in terms of the costs of manufacturing, then, we are facing significant environmental damage, adding to but probably not as great as the damage human activities are already making. This would undermine the persuasiveness of appeals to make sacrifices to preserve the ecology of the planet in its present, though damaged, state. But, to repeat, for many people this would not be overwhelming if there was no alternative. It would be enough to doom the project for those who weigh the health of the planet above the survival of any one species. Ask yourself which side you are on.

One very uncertain and worrying environmental aspect concerns the effect on the earth's atmosphere of the rocket launches. Remember that the Musk plan involved 10,000 spaceship launches. The danger is not carbon dioxide – and rocket engines use less fuel per pound lifted than passenger planes – but soot. Rockets release much more soot in relation to their fuel use than airplanes, and soot persists in the atmosphere for a long time and has a large warming effect on the atmosphere. Moreover, much of the soot would be in the upper atmosphere, where pollution has not been much studied. Rocket engines also damage the ozone layer, which protects life from ultraviolet radiation. Some consider this a greater danger than the soot.

These dangers have not been much researched, and they will vary with different types of rocket using different fuels. Solid fuel rockets, which are the vehicle of choice for NASA, though not for Musk, do more damage than liquid fuel rockets. The most effective liquid fuel in use is UDMH, which is very toxic and very prone to accidents. Hydrazine is sometimes used, and is very polluting. Musk's present rockets use kerosene – and oxygen – while the rockets he plans for a Mars trip would use methane. The threat is real enough, though, to suggest that much more study is needed before we pay enormous amounts for a chance to perform actions that might inflict irreparable harm.

A delicate issue is how to control private projects which present untested environmental dangers. Perhaps I am being too pessimistic about the dangers; perhaps the colonization enthusiasts are being too optimistic. But a sensible minimal requirement handles both of these. Any organization proposing a project should submit its plans to the community of environmental scientists and climate modellers, and an independent judgement should be made of their conclusions.

Opportunity costs

We should ask what opportunities we would miss by spending large amounts on Mars colonization and other space travel projects. What else could we get for these amounts? Where to begin? This amount of money could provide education for large numbers of people, raising them out of poverty or towards the kinds of jobs that a modern economy needs. It could then foster a greater world GDP, which could fund better-supported projects, including non-colonizing space projects. It could fund research on sustainable food supplies. It could fund the replacement of the fossil fuels that are driving global warming with sustainable sources of energy. It could provide clean drinking water and relief from endemic disease for much of the world's population. (The cost of eliminating malaria has been estimated at $8.5 billion.) It could provide the research and the hardware needed for protection against rogue asteroids. With a bit of imagination it could be used in campaigns to stem population increase or to alleviate the threat of global disorders brought by climate change and over-population. It could fund the setting aside of areas of the planet, such as the Amazon, for undisturbed nature. (The eminent biologist E. O. Wilson has suggested that half of Earth be set aside. Don't expect it tomorrow.) There has been recent progress in techniques for refor-esting deserts; it would be good for human and other life to apply these on a large scale.

Any one of these could be tackled with this amount of money. We could make progress on all of them, and all of them would spur technological and social progress of other kinds. The world is rich enough that a powerful world dictator could extract enough money from the population for a Mars colony and for these other needs. In a recent interview Musk suggests that "1%" of resources directed at improving conditions on Earth should be directed at a Mars colony. I am not sure what resources this is 1% of. That proportion of the earth's GDP would certainly be enough for his aims, though 1% of the money that is actually allocated to improving the ecology of the earth, and the situation of its worst-off, probably would not be. But that is not how things work in the world as it is. So, realisti-cally, we must decide. There are very few dangers, and quite definite benefits, to these alternative uses of global wealth.

These have been uses simply for the Mars project money. But the project would make rather specialized resource demands. It would need specialized personnel – engineers, experts in space travel (rocket science!) – and specialized manufacturing capacities. No doubt there are many valuable terrestrial projects that could use these particular resources, but two seem particularly important. Both focus on real rather than merely cosmetic environmental progress. One is the construction of transport networks which would reduce the use of private cars and airplanes. (Electric vehicles still place resource demands, particularly in the manufacture of batteries.) The other is designing and constructing systems for storing solar and wind energy, so that they can be used during times of darkness or calm. Vast power grids, both east–west and north–south, would allow much better use of intermittent sources of power. But designing and constructing them would require engineering and scientific sophistication.

The greatest lost opportunity, as far as I'm concerned, is from disrupting the desperately needed transition to a post-industrial economy on Earth. We cannot sustain the level of industrial production by current means without doing irreparable damage to the planet in terms of harm to the climate, the oceans, fresh water, and the

array of life Earth presently contains. Moreover, if the
current level of rich-country consumption is applied
globally, and achieved by present means, vastly more
damage can be expected. So we have to find a path to
different and less damaging ways of making things and
raising standards of living. This will take a while, but
a century or so of reduced industrial output is a small
price to pay. (This is a mere moment in human history,
let alone the history of life on this planet.) But entering
into a new and demanding industry, the manufacture
of unprecedentedly large and powerful rockets and the
spacecraft they power, would block this process.

Who makes the decisions, and how?

The costs are high, and the stakes are high. If the
dangers are likely, then very high costs are easier to
justify. Given the political difficulties of funding such
expensive projects with public money, there is a strong
case for leaving them to private or commercial initiative, whether driven by profit or by conviction. But if
the lives and the environment of all of us are at risk,
then we all – all people, and not just all those connected
with whatever organization is paying the bills – can

demand the right to be consulted. There is also the question not just of who decides but how they do it, what decision-making procedures are used.

Decisions can be made by default, and then we find ourselves having to live with a decision that someone else has made. On our topic we must look out for commercial interests, government bodies, the general lobby of space enthusiasts, and those with a vested interest. (As the novelist Upton Sinclair said, "It is difficult to get a man to understand something, when his salary depends upon his not understanding it.") Not that any of these need to be illegitimate or malicious. Some things are best left to those who are knowledgeable and committed. And some are not. The most important thing is that, if those most affected do not make the decisions, they are informed and willing to hand them over to others.

There is a lot of politics lurking here, since it may be a struggle to get fair decision-making styles applied. Think first of projects threatening high danger to the general public: a sizeable probability of consequences that would affect everyone for the worse. The bad effects might be serious environmental damage, great social disruption, prevention of more desirable projects, and even threats to the survival of human life on Earth. Then principles requiring social risk avoidance apply. That is, the public should insist that those who want the project to go ahead should clarify the dangers and make a case that the possible benefits outweigh the possible harms. The burden of proof would be on those who want to proceed. One important instance of this concerns environmental damage. Organizations that propose ambitious projects with even a slight danger of environmental damage should submit their plans to the ecological and climate-modelling scientific communities and not proceed unless a consensus emerges that the damage would be slight and reversible. Perhaps they should post damage-compensating bonds to avoid the spectre of harm followed by bankruptcy.

It is easy to say that this would be the way to decide extreme cases. And it obviously would be; everyone's children would be affected. But making the decision to go this way would not be trivial. Political action and public protest might well be needed. The most acute conflict would happen if, on the one side, we had leaders of industries who saw potential profits or satisfaction of their personal interests and, on the opposing side, there were numbers of non-influential people and those mobilizing them. It would be in everyone's interest if a political process could defuse the tension. But this would be made harder by the difficulty of enforcing the decisions of any political body with a world jurisdiction.

A second category is very delicate. Suppose we have serious but not overwhelming danger to the public, in terms of life, welfare, and environment. Public involvement would be appropriate, but the criteria for a decision are more fluid. In these middle cases it makes sense to weigh costs against benefits, though the costs are likely to be shared by many people and the benefits by fewer. Such weighing is never uncontroversial, and it is very sensitive to precise estimates of the benefits, losses, and probabilities. One rather conservative proposal is that those in favour of colonization should propose such estimates and the larger public should have a say in whether these proposals are reasonable. However, there is the small problem of turning this "should" into "will"!

In a third category the stakes are lower. The danger to the public is not great and the benefits to those undertaking the project are significant. The obvious procedure then is that the decisions be commercial: if you can do it and you can compensate those who are harmed, then go ahead. Someone has to be convinced that the project does fall into this category, though, and that the compensation is guaranteed. An appropriate and convenient body to be convinced is a national or supranational government with relevant jurisdiction.

When the costs are significant it is important to be clear about who is to pay them. This is something to keep in mind especially when opportunity costs are involved, as they are easy to ignore, and worth emphasizing especially when the proponents are largely well-off inhabitants of advanced economies. (No one should be left able to say "They have ruined our planet so that their children have an escape.") Costs should be distributed fairly and those least well-off should be least burdened. That is a classic definition of a just political system.

FIVE

Colonization without Humans

Human colonization of other planets faces many problems. For that matter, all travel in space by humans faces many problems. For my money, the prudent course is to delay and think more carefully what we can do, what we want to do, and what the costs would be. And, crucially, what the right timetable is both for developing the technology and for settling our priorities. One major argument against delay is the suggestion that the human species faces likely extinction on this planet, as described in chapter 1 and doubted in chapter 3, so that we should find a refuge as soon as we can.

If action in the next decades is not a priority, then we should be reflecting on really long-term plans and on what we value in them. That is the purpose of this chapter. A theme of my criticism of plans for colonization in the near future has been their hopeful fantasy quality, taking ideas from science fiction and trying to give them scientific substance while ignoring unwelcome facts. But in this chapter my discussion has some of that same character. The central question is what we want to aim for in the long run. That means reflecting on what we most value, and the possibilities I describe here are meant to prompt such reflection as much as to describe present possibilities that we know we can achieve. Are we considering all of our options?

Human ends

You are going to die. And if you are more than eighteen years old the knowledge that your life is finite is something you take in your stride, though you would have the end come later rather than sooner. The human species is also going to end, sooner or later. I think that a mature attitude to this inevitable fact is like a mature attitude to one's own death, and for very similar reasons. First, briefly, these reasons.

Many philosophers have argued that living forever would be a burden rather than a blessing. Adults are typically more troubled by the deaths of people they love, particularly their children and grandchildren, than by the prospect of their own deaths. This is reasonable. For among the many things we care about are the welfare of people close to us, the welfare of large numbers of people we do not know, and the balance of success and failure in our own lives. We care about so many things that the question as to whether we live to be 85 or 110 does not appear as a top priority.

Many people, I suspect most people, would prefer a gentle anticipation of their death than dying somewhat later in pain and distress. Most people would shorten their lives to lengthen those of others they care about and to advance causes they care about. Most people are comforted by the fact that others will continue after them and address these cares. When we reflect, we find our projects and those of people we care about are more important.

Now to the end of the species, a kind of death. It is inevitable that, as the sun ages, it will first become so hot that life on Earth will be extinguished; the sun will then expand so that it actually occupies the space where the earth is. (Do not cancel your insurance; this will not happen for five billion years.) But humans will surely not be around anywhere near that long. War, famine, disease, or ecological destruction will have done the job first. But these pale in comparison with the inevitability of another factor, evolution – or even old-fashioned natural selection. Modern humans have been around for about 250,000 years, following a much longer period during which pre-humans slowly changed to fill the niche we now occupy. Similarly, someone watching for a couple of hundred years, beginning now, would not remark on decisive species-distinguishing changes, yet in 250,000 years many small changes will have added up to something very new. It is like watching the hour hand move.

But it will happen much more quickly than this. We can now influence our own genetic makeup. Biotechnology aided by artificial intelligence will be irresistible. We are already on the brink of using CRISPR and other gene-editing techniques to treat hereditary diseases, in effect modifying genetic inheritance. One likely pattern is that a few parents will take the option of giving their children unusual intelligence and social skill, and then the following generation will have the choice of either doing the same or producing children who will be outcompeted by the changelings and will make a smaller contribution to the gene pool. In a couple of hundred years we could see changes as great as, say, that between *Homo sapiens* and *Australopithecus*.

Either the slow Darwinian route or the fast biotech route will lead to the end of the species. Is this sad? Not at all. Others will continue the concerns of humanity. They just will not be others with whom we personally can mate. (Well, much of present humanity wouldn't even consider mating with me.) So the end of this particular species is much like the end of this particular individual: not to be prevented, not to be whined about, but presenting opportunities to be grasped. There are developments to be avoided, to be sure. One of them is that the successors to humanity might have repugnant inhuman aims. Suppose that they were even more committed to warfare than humans are, for example. They would not last long in that case. Or suppose that they had no concern for the ecology of the planet and species besides their own. These for me would be reasons to resist the development. But it is really a political issue: we must make sure that the development of a successor species is shaped by all of us, and not just by a biotechnological elite.

There is a more drastic possibility. Many think it is a likelihood. The biological human species might be succeeded not by a biological species as we know it but by very different creatures. They might be purely artificial intelligences, or they might be cyborgs, mixtures of the organic and the inorganic, or they might be artificial life with a carbon base but using unprecedented genetic mechanisms. We should have the same concerns about these. Will they advance the projects that are important to us? What will their attitude be to old-fashioned life?

But these possibilities also open up opportunities. If we can design creatures that function on this planet better than we do, then we have a chance of designing creatures that function elsewhere in the universe better than we would. This undercuts the motive for escaping our fate on this planet even further. For we can make sure that life goes on and the things we care about are continued, even if not by us.

If our activities result in colonies on other planets, then we have caused those planets to be colonized. We have colonized them. (Just as if I order an accomplice to put poison in a victim's drink I have poisoned them. Sometimes other agents are the tools you use to do something.) So, in the futures I am describing, human beings do colonize other planets, but indirectly.

What we care about

It troubles each of us much less that others will live and replace us when we know that they will advance projects that matter to us. It should trouble us much less that the human species will have a successor if we know that it will advance projects that we value. What are these projects?

What you value is not the same as what you want. For one thing, you may take some of your desires as quirks of personality or temporary preferences which you would not mind losing. On the other hand, you would work to make yourself want, or continue to want, things that you value or care about. So while you take your fondness for lavender ice cream to be a peculiarity that your successors may shrug their shoulders about, if those who follow you do nothing to encourage tolerance, love, or curiosity, you will find it tragic that they are the ones who will follow you. It does not matter for present purposes whether these are simply marks of the distinction or at its heart. They might be features of human psychology, though fundamental ones, or they might be one way we grasp what is really valuable. Either way, they point to something basic about us, that there are things we will try hard to hold on to.

An important feature of values, as opposed to whims, urges, and obsessions, is that it is much less important to us whether we realize them ourselves or whether someone else does. (Consider two friends who share two important tasks: they each choose to focus on the one they are best suited for.) The most important thing is that the value be acted on by someone. That is one reason why people leave money to worthy causes.

Things that we value tend to focus on making: planning and creating families, societies, works of art, science, and all the other human works that we bring deliberately into existence. I shall take it that works that stand up to criticism over a period of time are central among the things that we value and for which we value others. This leaves a lot of room for variation. Some will most value the works of culturally recent people – the art, the political systems, the intellectual accomplishments – all occupying a tiny blink of human history. Some will most value elements of human life going back much further: the families, the generations-long conversations, the results of loyalty and altruism. And some will most value the results of our aliveness: the creation of more life, and handing it on to other generations, the search for new sources of pleasure and satisfaction. These overlap, but it makes a difference which are most vivid for you.

The distinction between wants and values bears on a frequent remark of Mars colony enthusiasts, including Elon Musk. It is that they want to see a colony within their lifetimes. One could describe this as selfish, preferring their own satisfaction at the expense of the quality of the accomplishments. I would rather describe it as a confusion between what one would like to be accomplished for its own sake and what one would like to experience. It is certainly good to do and to see what you deeply want to do and see, but any reasonable person will weigh this against things happening that they equally deeply want to happen.

Imagine that you have a child who is much more intelligent than you. She becomes a composer and writes highly original pieces that are widely enjoyed by both musicians and the general public, and which inspire mathematicians with their rigorous creative structures. You would surely be proud of her, and part of your pride would be that her capacities are so much beyond yours. The same would be true if she was a niece or a distant cousin. You would shorten your life to advance hers and take a little credit for your small role in bringing all this wonderful music to the world.

Push the story a little further. Your daughter is born with a genetic problem which will give her constant pain unless she receives gene therapy. It will prevent her from developing or exercising her talents. The therapy is given in a form that will more than compensate for her problems. The result is the brilliant musician. Again you would feel proud for the small role you played in what is valuable about her. She could be valuable in many ways: instead of being a brilliant musician she might be exceptionally warm and altruistic. So the intervention of a little technology, in this case genetic technology, does not change things. You are glad to be part of a story that advances things you value. So you would have no objection if a child of yours surpassed you partly as a result of genetic intervention. Similarly, you should not object to future beings, children of the present human species, who surpass us partly because of genetic intervention. We should be glad to see that they can do valuable things that we cannot.

One thing a descendant species could be better at is spaceflight. It might be capable of long periods of suspended animation. It might be more resistant to radiation. It might be more suited to interface with machines. We, present-day humans, are not suited for life beyond Earth; some of our descendants might be. If spaceflight is among our ambitions, we should welcome the arrival of our better adapted successors.

Seen this way, wanting to preserve the human race by colonizing other planets with present-day human beings is like the hopeless aim of gaining personal immortality by making pyramids or empires. It won't work.

The coming machines

Machines controlled by computers instantiating artificial intelligence are going to perform more and more functions. That is uncontroversial. Some think that this points to a very subsidiary role for humans, and perhaps to the end of the species. I am far from convinced. But there are many human and machine scenarios. Consider just one. We program computers to design computers and write software for them which designs other computers and writes software for them, and the eventual result is machines that we cannot begin to understand and which show all the signs of conscious and extraordinarily intelligent goal-directed action. Some of these machines are used in space exploration to replace humans. To do this, they are programmed to act autonomously to achieve scientific aims. And for the sake of these aims they establish robotic bases on Mars and on satellites of Jupiter and Saturn, deliberately excluding humans from these activities because humans do them so badly. (See the remarks at the end of chapter 1 about the growing capacity of computation to replace humans in space exploration.)

Supposing that the machines are in fact better at getting and analysing information than humans are, we would find we had little choice but to embrace their oversight of the process. (Robotic spacecraft already collect nearly all the information we get from space. Human astronauts undertake missions just for reasons of national prestige and as preparation for an imagined future.) The alternative would be to proceed without them and accept that the conclusions we formed were fragile and not supported by the possible evidence. But that would undermine the whole project of science. The situation is essentially like the use of computers in mathematical proof even now, or like computer analysis of data in complex experiments in physics, where, if we refuse cybernetic help, our scientific ambitions are limited.

A very telling consideration here is whether this process, essential parts of which would be beyond human understanding, provides theories and explanations that humans can grasp. They would be like the versions of physics and astronomy that we give to children. ("Once upon a time the universe began with a big bang and spread outward from there," "electrons go round and round the nuclei of atoms in circles.") If we were given these "childish" versions, then we would understand the universe better than we would without the help of artificial intelligence. If they were not given to us, then we would be reduced to the status of toddlers ourselves. Our physics would be like chess as played by merely human players. On the other hand, we would know that real physics was being performed, though not by us.

Suppose that few of the projects that human beings are proud of are best carried out by human beings: machines produce better music, better mathematics, and better science and manage the affairs of Earth better than we could. This would prompt a major reconsideration of humanity's conception of itself. Over many years we would revert to being what we once were, an animal among others. This need not be sad: the animal pleasures would still be ours and we would know that the planet was in good hands, much as domestic animals trust humans to take care of things, letting them provide green pastures and still waters. In fact, characteristically human life would proceed largely unchanged. Art, love, wonder. It would be just as well not to take this as something sad, for we wouldn't have much choice in the matter.

That is a not-very-good-case scenario, though. At least as likely is the possibility of symbiosis. We stop doing things that the machines do better, and both their tasks and ours go better when we do the things that we do better. The things that we do better, as it turns out in this scenario, include scientific and artistic creativity, the formulation of novel long-term projects, and transferring techniques from one domain to another. The machines are wonderful, for example, at brute force proofs of mathematical theorems, but they are pretty feeble at indirect oblique surprisingly roundabout proofs, as when you prove something about prime numbers by using complex analysis. They are wonderful at producing imitations of paintings by established artists but pretty feeble at writing plays in which people make subtle jokes which seem to come out of nowhere. They are unbeatable playing chess but

hopeless at inventing games that are satisfying challenges for either people or machines.

This is just a possible development, but it would
not be surprising. It is what evolution would suggest.
We humans are adapted to changing and unpredictable
environments where getting a good-enough approxi-
mate solution is a life-or-death matter. Our history
has focused less on exact solutions to variations on
known problems. We make machines for that. Or, to
put it differently, if we emphasize biology and the long
term as much as physics and the short term, then the
transience of this particular moment for this particular
species fades in importance compared to life in general
and what life can accomplish.

Extended colonization 1: seeding

So we should not fear that other species may come after
us, especially if they are other species that will continue
what we find important. What has that got to do with
colonization projects? One connection is immediate.
The motive for establishing a foothold for present-day
humans beyond Earth is undermined (though it is far
from sure to work, anyway). As long as intelligent life
goes on here, we will be succeeded by others as we

would be by our children. And if our strand of intelligent life does not go on, then sooner or later life will sprout intelligence, so the delay is just a delay. The biggest peril, on this point of view, is the extinction of all life, all higher life. That is what we must most defend against, and the defence must happen on Earth.

But variations on the classic colonialist idea are still on the table. Given that the bulk of activities beyond Earth are best left to machines, there is a variety of ways in which we can share the project with them. The first is the project of seeding life elsewhere.

Whether or not life on Earth is actually the result of life-like processes elsewhere, as is sometimes claimed, it could have been, and this raises the possibility that life from Earth could spread elsewhere. Humans, on behalf of earthly life, could "colonize" other planets with bacteria and other simple organisms by sending cell cultures by spacecraft. (We try very hard at the moment *not* to do this.) Radiation damage is a threat to travelling microorganisms so the packaging would be essential. But it would not have to be bulky.

Conditions on another planet would not be hospitable to organisms that have evolved to live on Earth. Organisms specifically designed for the conditions they would be plunked into would be better. Their basic chemistry and cellular design would have heat, cold, freezing water, methane, lack of oxygen, or whatever, in mind. The aim would be to develop organisms that would flourish, evolve, and transform a particular alien environment. The host planet might become more like Earth. But it might not; it might become simply more hospitable to the descendants of those immigrants.

We are only beginning to understand how to design organisms, but a beginning has been made, for example in the work of Craig Venter. It will be a very active field of research in the next decades. For the present purpose, seeding life elsewhere, it would need to be combined with a novel field which would explore the suitability of possible life forms for various conditions. There would be experiments duplicating alien environments on Earth, or relatively nearby, to see which organisms coped with them. There would also have to be intensive discussion of the ethical question as to what life forms we wanted to encourage: some might out-compete organisms already in the new environment, some might not be promising possibilities for life in general, and in the long run some might out-compete *us*.

An analogy: when animal life migrated from the oceans to the land it exploited features that had already evolved in water. But more evolution resulted in creatures intrinsically adapted to life on the dry surface of the earth that cannot survive underwater. Similarly, we can expect that life that is adapted to interplanetary travel and to flourishing on other planets will be different from Earth life in basic ways. It is likely that it would not do well if returned to Earth.

Seeding life further afield, elsewhere in the galaxy, has to proceed differently. It would be impractical to have probes visit candidate locations and report back so that further automated expeditions can transport suitable life. For one thing, there would be an enormous time delay. For another, many candidate locations will prove on closer examination not to be suitable, so there would have to be an enormous number of probes. These two factors combine, since it is only with the very nearest stars that the time and travel delay is manageable. We are a long way, perhaps centuries, away from being able to do anything of the kind. But we can now imagine factors that might turn out to be relevant. Nanotechnology would supply tiny probes that would blanket the galaxy, much like the reproductive strategies of organisms such as dandelions. (Spamming the universe.) They would then assess the conditions at any destination and, where possible, synthesize suitable biochemistry.

By the time any of this becomes possible better ways may also be available. We need all the time we can get, starting now, to consider whether it is something we want to do and, more subtly, how important to us it is.

Present-day unmodified humans could even be colonists, though temporary ones. Astronaut-scientists might be a useful addition to planetary bases at the centre of the non-human colonization. Being human, they would not be suitable for long-term stays and would be rotated out periodically. Still, their contribution might be an essential stopgap both in establishing colonies and in extraterrestrial science. And, although I am sure that the commercial exploitation of space will be largely robotic, space-faring humans can play a role in this too. Primarily non-human colonies and their protective machines might be symbiotic with these activities.

Extended colonization 2: surrogate experience

We might be able to stay on Earth and participate in colonies elsewhere. This could happen in several ways. The most basic is virtual reality, where the human mind is linked to an extraterrestrial robot. The human would experience sensations collected by the robot and would help determine its actions. The time delays of communication with a remote agent would inhibit simple-minded ways of implementing this over any appreciable distance. So we would need ways of sending and receiving stretches of data, not synchronized to one another, and the mechanics of this would surely vary from one project to another. The simplest project would be a kind of tourism, where the robot was largely autonomous and the distant human set general intentions – "explore this crater" – and left the implementation of them to the robot while eventually receiving its experiences. Less simple projects, particularly those involving danger or scientific research, would need subtler arrangements. One big unknown here is the extent to which human capacities can complement those of advanced AI. Paradoxically, if, as I suspect, the peculiar human brain can add something that is hard to reproduce inorganically, then the demands for close interaction between human and robot are greater, and time delays and the like are more of an obstacle. The more irreplaceable we are, the less the prospects for ersatz exploration. Here too a lot depends both on the nature of the task and on the specifics of how human beings or their descendants will best interact with artificial intelligence of the future. And we just do not know enough about this: we probably will not until we get there.

We can get a stronger form of human participation if we assume something completely unproven, on a par with the science-fictional fantasies that I have often tried to undermine. This is that our minds can be uploaded into computer memory, a staple of science fiction and superficially imaginable. It requires information from a human brain to be transferred to another medium, such as a futuristic version of a contemporary computer, so that memory is available, personality is intact, and consciousness can continue as before. Before we can say that this is possible, let alone practical, we would have to know more about the physical basis of consciousness, the workings of the human brain, and what makes one experience the same as another. You aren't going to get answers to these from me here!

If human consciousness can be uploaded to non-human forms, then people can come along for the ride. They can be small and light; they won't need oxygen; they can be turned on and off as convenient. They can even return and tell their adventures. Return is not essential, though, because destruction of uploaded memories and consciousness need not mean destroying the original bodily instances. So the uploaded mind could experience the adventure, transmit it back to its original, and then accept its fate. This possibility might make us less cautious about taking risks with the bearers of the human mind. (Perhaps we have already been visited by this technique, with tiny visitors cleverly disguised as fruit flies.) There might be situations when the expedition was more successful with this addition. It *might* even do a better job of spreading biological life more widely.

Either in the virtual-reality form or in the uploading form, we would get new adventures on Earth. Miniaturization would be the secret to many of these. One avatar might be a small, perhaps tiny, model human equipped with sensors. Then crossing a lake would be an ocean voyage, and crossing a field would be a struggle with monsters; birds would be dragons, rainstorms hurricanes, and fields wild continents. And, more to the present theme, circling the earth would be voyaging the galaxy.

We must appreciate how open the future is, how many vital developments we will not have anticipated until they are upon us. That ought to be a given of futurology, as suggested by its patchy past. We need contingency plans; we need to think out the technology and basic science that might be relevant; we need to buy time to investigate properly. And, too often neglected, we need to ask ourselves hard and sustained questions about what it matters most to us to achieve.

Why Human Colonization is a Bad Idea

I have been describing the pros and cons of colonizing other planets, especially Mars, with present-day humans, with emphasis on the cons, since that is the side that is neglected at the moment in public discourse. There are some very general considerations behind many of the particular points. They are broadly biological and social in character, and they contrast with the physical science that dominates many discussions of the topic.

Besides defending some broad positions, behind the detailed pros and cons, I have also raised some large questions which are not usually put in the foreground of issues about planetary colonization. While I would be delighted if after careful consideration you conclude that my positions are absolutely correct, it is at least as important that we ask the right questions as we continue to think about the issues. So I begin with the two central questions.

The first question is how we should weigh the continuation of the present-day human species against (a) the continuation of biological life and (b) the continuation and increase of intelligence. These are not the same. Biological life might continue without harbouring the most powerful possible intelligence. And intelligence might develop in the absence of continuing biological life. Our on-the-spot preferences are much less important here than what we think after reflection and debate.

The other central question is how long we should think about colonization proposals before doing anything. On the one hand, delay for more careful planning and for advances in technology has obvious advantages. On the other hand, there may be risks in delaying too long. This is best thought about in the context of a long-term vision of the future of technology and human exploits. Such visions are inevitably vague and tentative, but year-to-year (or century-to-century) plans make more sense against such a backdrop.

Now my definite conclusions: you may not share them all, so these may be subjects for re-reading and dispute.

We are not well adapted for life away from Earth. Humans evolved to survive under particular conditions and need complex high-tech adaptations to live elsewhere. We need a lot of food, a specific atmosphere, a particular temperature range, and the absence of radiation. We surely do not know all the detailed features of the environment that we need in order to function. As a result many tasks are better performed by intelligent robots, whose capacities are constantly increasing. Some forms of non-human life, such as specially designed one-celled organisms, may be much more promising for surviving far away.

A technologically advanced civilization requires a large number of interacting people with complementary specialist knowledge. Only with cutting-edge technology can humans survive beyond Earth. This means that a self-sufficient human colony on another planet must be greater than a minimum size. It must be at least the size of a small self-sufficient technological nation. (Understanding the structures of human societies, and the thoughts and actions they lead to, is at least as challenging as physics.)

A remote colony does not give immunity to the threats to life on Earth. Participants in a terrestrial nuclear or biological war could as easily target a colony. Even a post-war Earth would be more hospitable than Mars or Venus. Social tensions that break human society would occur just as easily elsewhere. In fact, a colony would be more vulnerable to many threats than the population on Earth, because it could not survive without its specialized technology and its artificial living spaces. To these we must add the possibility that a colony elsewhere in the solar system might make existential dangers on Earth, from war or ecological destruction, more rather than less likely.

The place of humans will be taken, sooner or later, by other creatures. We have some choice of what these other creatures are, and it is important that we think through these choices. We value some aspects of our individual lives and of our species more than others, and clarity on this is vital to making decisions that we and our successors can live with.

Think of humanity as a particular species that has evolved to fill a particular ecological niche in a tiny fraction of the time that there has been life on Earth. Our intelligence and our imagination are rooted in our history and limited by it. There is a lot that we do not understand and many possibilities that we cannot get our minds around. Knowing this, we can at least resolve to make our plans slowly and carefully, giving ourselves enough time to resolve basic scientific problems about human possibilities beyond Earth. And we can think on a biological timescale, where species succeed one another over thousands or millions of years. We should take the long view.

Further information on the topics discussed in these chapters can be obtained from the references below. Much information can also be got from Internet searches on reputable sites such as Wikipedia and those listed in the next paragraph. I am not citing simple factual information that can be found in this way.

Websites of general relevance to the topic
of planetary colonization

www.nasa.gov/
www.hawking.org.uk/
www.spacex.com/
www.mars-one.com/
www.marssociety.org/
www.astronomy.com/magazine

On all of these, judicious use of the search function would be helpful. The most relevant items are often not displayed on the homepage.

Books and articles of general relevance

Musk, Elon (2017) "Making humans a multi-planetary species," *New Space* 5(2): 46–61 (https://doi.org/10.1089/space.2017.29009.emu).

Tyson, Neil deGrasse, Jeffrey Simons, and Charles Liu (eds) (2016) *StarTalk: Everything You Ever Need to Know about Space Travel, Sci-Fi, the Human Race, the Universe, and Beyond* (Washington, DC: National Geographic).

Chapter 1   Escape from Earth?

The Four Horsemen of the Apocalypse appear in the Book of Revelation, but as symbolic figures rather than names of disasters, so different lists name and order them differently.

For general information on the risk of nuclear war, see the *Bulletin of the Atomic Scientists*: http://thebulletin.org/. For reasons why it is no longer so clear why we have dodged nuclear war for seventy years, see "Nuclear endgame: the growing appeal of zero," *The Economist*, June 16, 2011.

For background information about viruses and pandemics, see Crawford, Dorothy H. (2011) *Viruses: A Very Short Introduction* (Oxford: Oxford University Press). For the immediate danger of a pandemic, see Wolfe, Nathan (2009) "Preventing the next pandemic," *Scientific American* 300(4): 76–81.

The NASA DART plan for preventing asteroid strikes is discussed at www.nasa.gov/feature/nasa-s-first-asteroid-deflection-mission-enters-next-design-phase.

Lagrange points are positions on the path of the earth (or another planet) which are so balanced with respect to the earth and the sun that they maintain a constant position with regard to them. They have been suggested as locations for artificial colonies. There have been several instrument-equipped expeditions to these points. A recent one is the European Space Agency's Gaia craft. See http://sci.esa.int/gaia/.

A digest of current knowledge about how human life continued unchanged for hundreds of thousands of years until the mixed blessing of the agricultural revolution is Lanchester, John (2017) "The case against civilization," *New Yorker*, September 12, 2017. Lanchester is reviewing Scott, James C. (2017) *Against the Grain: A Deep History of the Earliest States* (New Haven, CT: Yale University Press).

A typical assertion of the general danger of artificial intelligence may be found at www.npr.org/2017/07/18/537844706/elon-musk-artificial-intelligence-poses-existential-risk. The danger can be argued for in terms of the unpredictability of algorithms, discussed at www.technologyreview.com/s/604087/

thedarksecretattheheartofai/. A good example is the unmanageability of computer-based mathematical proof, as described in Lamb, Evelyn (2016) "Two-hundred-terabyte maths proof is largest ever," *Nature*, May 26, pp. 17–18. Recent developments in integrating human and computer proof are discussed in Rehmeyer, Julie (2013) "Voevodsky's mathematical revolution," *Scientific American*, October 1. The frame problem for artificial intelligence is actually a complex of related issues; see Shanahan, Murray, "The frame problem," *Stanford Encyclopedia of Philosophy* (spring 2016 edn), at https://plato.stanford.edu/archives/spr2016/entries/frame-problem/. AI professionals usually do not take the dangers as seriously as some non-specialists do. See Calo, Ryan (2017) "Artificial intelligence policy: a roadmap," *SSRN*, August 8 (https://ssrn.com/abstract=3015350); Sofge, Eric (2015) "Why artificial intelligence will not obliterate humanity," *Popular Science*, March 19 (www.popsci.com/why-artificial-intelligence-will-not-obliterate-humanity). The extreme version of the danger is summed up in the concept of The Singularity, originally formulated by the science-fiction writer Vernor Vinge. It refers to a speculative moment when machine intelligence passes a threshold where positive feedback makes it unstoppable. It is developed in argumentative form in Kurzweil, Ray (2005) *The Singularity is Near* (London: Penguin).

Ways of making decisions in the face of very bad possible outcomes are discussed in Bostrom, Nick (2002) "Existential risks: analyzing human extinction scenarios and related hazards," *Journal of Evolution and Technology* 9(1); O'Riordan, Timothy, and James Cameron (eds) (1994) *Interpreting the Precautionary Principle* (London: Routledge); and Sunstein, Cass R. (2009) *Laws of Fear: Beyond the Precautionary Principle* (Cambridge: Cambridge University Press).

Chapter 2   The Colony Solution

The idea of assuring the continuation of humanity by expanding into space is a very general one. It is discussed in generality in Burrows, William E. (2006) *The Survival Imperative* (New York: Forge Press). Such plans inevitably involve predictions of technological progress, which are pretty unreliable. A classic explanation of why they are unreliable, that there is no substitute for scientific discovery, is found in Popper, Karl (1957) *The Poverty of Historicism* (London: Routledge).

In the case of Mars, many relevant facts are detailed in the book by Tyson and others, listed under general reading above. On the issue of water, recent observations suggest that there may be more water beneath the surface in frozen form than earlier estimates. See Dundas, Colin M., and others (2018) "Exposed subsurface ice sheets in the Martian mid-latitudes," *Science* 359(6372): 199–201 (http://science.sciencemag.org/content/359/6372/199); Greshko, Michael (2018) "Huge water reserves found all over Mars," *National Geographic*, January 11 (https://news.nationalgeographic.com/2018/01/mars-buried-water-ice-subsurface-geology-astronauts-science/).

Science fiction often features Mars, for the same reasons that colonization plans do, but without the constraints of realism. An unusually realistic depiction, though conveniently ignoring some facts, is in Weir, Andy (2014) *The Martian: A Novel* (New York: Crown). It is discussed critically in Paoletta, Rae (2017) "Mars might not be the potato utopia we hoped," available at http://gizmodo.com/mars-might-not-be-the-potato-utopia-we-hoped-1796713752. An informative background on the literary imagination of Mars is Bilger, Burkhard (2013) "The Martian chroniclers: a new era in planetary exploration," *New Yorker*, April 22.

As I write, there are signs of a renewed interest in colonizing the moon, though it is too soon for detailed plans. One reason, I suspect, is that the obstacles to colonization on Mars, as described in chapter 3, are becoming clearer.

For general information about SpaceX and Mars One, see their websites mentioned in the general reading above, and also Musk's "Multi-planetary species" article listed there. Also helpful are three presentations and expositions by Musk, which describe recent plans but are somewhat short of technical details, perhaps in part to preserve proprietary information: an interview with Musk at www.youtube.com/playlist?list=PLsJnBOe4es TYVKhyconcb4qK915ensFMC; a polished presentation of recent plans at www.spacex.com/sites/spacex/files/making_life_multiplanetary-2017.pdf; much of this material is also in a conference presentation by Musk, at www.youtube.com/watch?v=tdUX3ypDVwI.

The MIT study of Mars One is most concerned with costs, so I list it in the reading for chapter 4.

For the NASA habitat-printing competition, see www.nasa.gov/directorates/spacetech/centennial_challenges/3DPHab/index.htm; www.space.com/37427 -nasa-mars-habitat-challenge-phase-two-winners.html; and www.nasa.gov/press-release/nasa-awards-400000-to-top-teams-at-second-phase-of-3d-printing-competition.

Terraforming is discussed at www.science20.com/robert_inventor/trouble_with_terraforming_mars-126407. See also "These experiments are building the case to terraform Mars," *Astronomy* (2016), June 14, 2016 (www.astronomy.com/news/2016/06/these-experiments-are-building-the-case-to-terraform-mars).

Two contrasting studies about how Mars life is threatened by the introduction of life from Earth are, first, a study that suggests that terrestrial microbes will have difficulty surviving on Mars: Schuerger, Andrew C., Richard Ulrich, Bonnie J. Berry, and Wayne L. Nicholson (2013) "Growth of Serratia liquefaciens under 7 mbar, 0°C, and CO-enriched anoxic atmospheres," *Astrobiology* 13(2): 115–31; second, a study of which bacteria best survive Martian conditions, which suggests ominously that some human fecal bacteria can survive: Schuerger, Andrew C., Doug W. Ming, and D. C. Golden (2017) "Biotoxicity of Mars soils: survival of Bacillus subtilis and Enterococcus faecalis in aqueous extracts derived from six Mars analog soils," *Icarus* 290: 215–23.

For the contrast between the views of deGrasse Tyson and Musk, see www.youtube.com/watch?v=lQd7zqyd_EM and http://aeroastro.mit.edu/videos/centennial-symposium-one-one-one-elon-musk.

On legal issues concerning space, see "Treaty on principles governing the activities of states in the exploration and use of outer space, including the moon and other celestial bodies," United Nations Office for Disarmament

Affairs (http://disarmament.un.org/treaties/t/outer_
space). Also a principled campaign that might at first
seem like whimsy, described at www.wnyc.org/story/
man-who-would-be-king-mars-podcast/.

Chapter 3   Problems with Colonies

The NASA DART plan for dealing with asteroids is
cited in the reading for chapter 1.

For background facts about genetic diversity, see
Gagneux, Pascal, and others (1999) "Mitochondrial
sequences show diverse evolutionary histories of African
hominoids," *Proceedings of the National Academy of Sciences* 96(9): 5077–82; and Frankham, Richard (2005)
"Genetics and extinction," *Biological Conservation*
126(2): 131–40.

For the danger of perchlorates in the Martian
soil, see Davila, Alfonso F., David Willson, John D.
Coates, and Christopher P. McKay (2013) "Perchlorate on Mars: a chemical hazard and a resource for
humans," *International Journal of Astrobiology* 12;
Warmflash, David (2016) "Salts on Mars could pose
unseen hazards to explorers," *Astronomy*, June 21
(www.astronomy.com/news/2016/06/salts-on-mars-
could-pose-unseen-hazards-to-explorers); and David,
Leonard (2013) "Toxic Mars: astronauts must deal with

perchlorate on the red planet," www.space.com/21554-mars-toxic-perchlorate-chemicals.html.

The complexity of technological society is a basic social fact that is often ignored in these discussions. For background, see Groenewegen, Peter (1987) "Division of labour," in *The New Palgrave: A Dictionary of Economics*, ed. John Eatwell, Murray Milgate, and Peter Newman (London: Macmillan).

3D printing beyond Earth is an important resource for planetary colonization, though it does depend on the availability of resources. See www.nytimes.com/video/science/space/100000005449852/3-d-printing-for-space-exploration.html.

Social issues need as much thought as technological issues. Possible social tensions in a Mars colony are discussed in Gohd, Chelsea, "A few reasons why living on Mars would suck," at www.outerplaces.com/science/item/16172-mars-colonization-life. Unanticipated social problems in a Mars colony, and tensions with Earth society, are a central theme of Kim Stanley Robinson's novel *Red Mars* (New York: Random House, 1993), the first volume of a trilogy whose later volumes fictionalize terraforming, among other topics.

Medical issues about spaceflight, including cancer, muscle wastage, and osteoporosis, are discussed in Buckey, Jay C., Jr. (2006) *Space Physiology* (Oxford:

Oxford University Press). Radiation dangers are an enormously worrying issue about which we do not know enough. For a NASA experiment, see https://spaceflight.nasa.gov/spacenews/factsheets/pdfs/radiation.pdf. For a popular article, see Kim, Gene, and Jessica Orwig, "Scientists overlooked a major problem with going to Mars – and they fear it could be a suicide mission," www.businessinsider.com/scientists-overlooked-high-radiation-mars-suicide-missions-2017-6. For a scientific article, see Cucinotta, Francis A., and Eliedonna Cacao (2017) "Non-targeted effects models predict significantly higher Mars mission cancer risk than targeted effects models," *Nature Scientific Reports* 7 (www.nature.com/articles/s41598-017-02087-3).

Experiments testing Mars habitats on Earth are discussed in Horton, Michael (2009) "Mars desert research station simulates Mars-like base," *Tech Fragments*, January 18 (https://techfragments.com/mars-desert-research-station-simulates-mars-like-base/) and Heinicke, Christiane (2018) "My year on 'Mars'," *Scientific American* 318, January 2 (www.scientificamerican.com/article/my-year-on-mars/) .

Risk compensation and moral hazard, where precautions actually increase risk, are discussed in Ecenbarger, William (2009) "Buckle up your seatbelt and

behave," www.smithsonianmag.com/science-nature/
buckle-up-your-seatbelt-and-behave-117182619/;
Lund, A. K., and P. Zador (1984) "Mandatory belt
use and driver risk taking," *Risk Analysis* 4: 41–53; and
Arrow, Kenneth (1963) "Uncertainty and the welfare
economics of medical care," *American Economic Review*
53(5): 941–73.

For more general considerations on how an extraterrestrial refuge might make for greater dangers on
earth, see Cave, Stephen, Lucianne Walkowicz, and
Huw Price (2016) "Colonise Mars as humanity's
plan B? It's a dangerous fantasy," *New Scientist*, April
1 (www.newscientist.com/article/2082833-colonise-
mars-as-humanitys-plan-b-its-a-dangerous-fantasy/)
and Walkowicz, Lucianne (2015) "Let's not use Mars
as a backup planet," www.ted.com/talks/lucianne_
walkowicz_let_s_not_use_mars_as_a_backup_planet/.

Chapter 4   Costs of Colonization

For how to discuss and model the costs of a complex
issue, see Broome, John (1993) *Counting the Costs of
Global Warming* (Cambridge: White Horse Press).

The MIT study of Mars One plans is Do, Sydney,
Andrew Owens, Koki Ho, Samuel Schreiner, and
Olivier de Weck (2016) "An independent assessment

of the technical feasibility of the Mars One mission plan – updated analysis," *Acta Astronautica* 120: 192–228.

Gerard 't Hooft's comments on Mars One are cited at www.theguardian.com/science/2015/feb/23/mars-one-plan-colonise-red-planet-unrealistic-leading-supporter.

The topic of environmental effects of colonization proposals is amazingly undeveloped. For some very basic considerations, see Burnham, Michael (2009) "Can space travel be environmentally friendly?" *Scientific American* 309; Moseman, Andrew (2009), "How to make a (more) environmentally friendly rocket fuel," www.popularmechanics.com/space/rockets/a6377/4330380/; Ross, Martin (2011) "Rocket soot emissions and climate change," www.aerospace.org/crosslinkmag/summer2011/rocket-soot-emissions-and-climate-change/; Inglis-Arkell, Esther (2011) "Will commercial space flight ruin the environment back on Earth?", https://io9.gizmodo.com/5727843/will-commercial-space-flight-ruin-the-enviroment-back-on-earth.

A general discussion of the vital issue of opportunity costs is Buchanan, James M. (2008) "Opportunity cost," *The New Palgrave Dictionary of Economics*, 2nd edn, ed. Steven N. Durlauf and Lawrence E. Blume (London: MacMillan). An example of a significant gain that could be had for a relatively small cost is that of eliminating malaria. See Zelman, Brittany, Anthony Kiszewski, Chris Cotter, and Jenny Liu (2014) "Costs of eliminating malaria and the impact of the Global Fund in 34 countries," PLoS ONE 9(12): e115714 (https://doi.org/10.1371/journal.pone.0115714); and *Action and Investment to Defeat Malaria 2016–2030*, www.rollbackmalaria.org/files/files/aim/RBM_AIM_ Report_A4_EN-Sept2015.pdf.

The interview with Musk where he speaks of 1 percent of resources providing a Mars colony is at http://aeroastro.mit.edu/videos/centennial-symposium -one-one-one-elon-musk.

Environmental suggestions often have hidden environmental costs. See, for example, this discussion of naïve avoidance of fossil fuels: Hawkins, Troy R., Bhawna Singh, Guillaume Majeau-Bettez, and Anders Hammer Strømman (2013) "Comparative environmental life cycle assessment of conventional and electric vehicles," *Journal of Industrial Ecology* 17(1): 53–64. As a result we often have to balance costs, benefits, and probabilities that are very hard to estimate. This is the topic of decision-making with extreme but vague information discussed in the last set of references to chapter 1.

Since colonization projects distribute costs and benefits unevenly, they raise issues of fairness. The idea that social proposals should be evaluated in terms of their benefits to the worst is a classic idea defended in Rawls, John (1999) *A Theory of Justice* (rev. edn, Cambridge, MA: Harvard University Press).

Chapter 5   Colonization without Humans

Discussions of the rationality of fearing one's own individual death go back a long way in philosophy. Contemporary discussions are Glover, Jonathan (1977) *Causing Death and Saving Lives* (Harmondsworth: Penguin) and Luper, Steven (ed.) (2014) *The Cambridge Companion to Life and Death* (Cambridge: Cambridge University Press). Discussions of the rationality of fearing the end of the human species are less common. One is found in Scheffler, Samuel (2013) *Death and the Afterlife* (Oxford: Oxford University Press).

The relevance of future biology to issues about colonization and spaceflight is the theme of this book. A basic genetic technology that will have major consequences is CRISPR. See Ralston, Geoff (2017) "Hacking DNA," https://blog.ycombinator.com/hacking-dna-the-story-of-crispr-ken-thompson-and-the-gene-drive/. Cyborgs, the fusion of biological and artificial life, are a long way off. But there are hints and anticipations in present-day developments. See "How brains and machines can be made to work together," *The Economist*, January 6, 2018, Technology Quarterly (www.economist.com/news/technology-quarterly/21733196-brain-computer-interfaces-sound-stuff-science-fiction-andrew-palmer). One of the contemporary figures who will have a great influence on future biotechnology is Craig Venter. For more about him, see www.jcvi.org/cms/home/.

Understanding the difference between desires and values can go some way towards making issues about moral choice less mysterious. See Frankfurt, Harry (1971) "Freedom of the will and the concept of a person," *Journal of Philosophy* 68: 5–20; and Smith, Michael, David Lewis, and Mark Johnston (1989) "Dispositional theories of value," *Proceedings of the Aristotelian Society*, Supplementary Vol. 63: 113–37.

It is a scientifically well-established conjecture (definitely not a scientifically well-established fact) that life on Earth might be the result of life-like processes elsewhere. Organic molecules are not uncommon in outer space. And one origin for them might be the fact that, as generations of stars go through cycles of development, explosion, and re-formation, atoms of substances further down the periodic table are created.

The universe is now thought to be some 13 billion years old and is expected to be producing stars for more than a thousand times that interval. Moreover, the type of star that typically develops into a supernova has a life of some 30 million years. So the universe is relatively young, there has been time for several cycles already, and there is time for even more in the future. The theoretical possibility of life having evolved in the past near stars that no longer exist, and speeding the evolution of life in its future, is quite real. See Warmflash, David, and Benjamin Weiss (2005) "Did life come from another world?" *Scientific American* 293(5): 64–71 (www.nature.com/scientificamerican/journal/v293/n5/full/scientificamerican1105-64.html). Possible cosmic origins suggest possible cosmic continuations, thus the idea of seeding. See Mautner, Michael N. (2000) *Seeding the Universe with Life: Securing Our Cosmological Future* (Washington, DC: Legacy Books) and Overbye, Dennis (2016) "Reaching for the stars, across 4.37 light-years," www.nytimes.com/2016/04/13/science/alpha-centauri-breakthrough-starshot-yuri-milner-stephen-hawking.html.

Another futuristic aspect to the relation between minds and machines is the possibility of uploading consciousness into computational machinery. For a mixture of views about this, see Chalmers, David (2003) *The*

Conscious Mind: In Search of a Fundamental Theory. (Oxford: Oxford University Press); Kurzweil, Raymond (2000) "Live forever – uploading the human brain: closer than you think," *Psychology Today* 33(1): 66–71, also widely available on the web; Rosenthal, David M. (2006) *Consciousness and Mind* (New York: Oxford University Press); and Dennett, D. C. (1980) "Where am I?" in *Brainstorms* (Cambridge, MA: MIT Press).

Both predictions about the future and scepticism about such predictions are important to projects such as this. In chapter 1, I cited Karl Popper's scepticism, and there is much in this chapter that he would be sceptical about. So it is important to see how fallible futurology can be. To illustrate the patchy track record of attempts to predict more than the immediate future, consider some examples from film: in *2001: A Space Odyssey*, there are no small computers but there are atomic rockets; in *Blade Runner*, set in 2019, there are intelligent robots but there are no cell phones. The pattern is that we anticipate improvements of present technology and science but, unsurprisingly, miss scene-changing discoveries.

Apocalypse, Four Horsemen of 2, 112
artificial intelligence 11–19, 53, 81–9, 96
  as complementing humanity 90–1, 98
  as taking the place of humanity 87–91
asteroids, dangers of 4–8, 26, 46, 52, 70

biotechnology 3, 81–2

Cassini 19
catastrophes, for a colony 46–52
climate modelling 31
commercial exploitation of space 37, 73–6, 95
consciousness, in machine form 97, 119–20
contamination of Mars by earth organisms 35
cosmic rays 7, 25
costs
  environmental 55–7, 66–70, 115–16
  monetary 58–60, 62–71
  opportunity 70–3
CRISPR 81, 117
cyborgs 82

DART 7, 105
death
  of individual humans 79–80
  of the human species 81–2, 117
decision-making 18–19, 39, 49, 73–6, 107, 116
descendant species from humanity 86–7

environmental costs 55–7, 66–70, 115–16
evolution 13–14, 80–1, 91, 93

food, on Mars 29–31, 41, 43, 52, 58, 60, 61–2
frame problem 13, 106

genetic diversity 42, 42, 50, 58, 111

habitats 229, 30–3, 34, 42, 51–2, 60, 109, 113
Hawking, Stephen 12, 18
human life, as historically recent 9, 80, 102
humans, unsuitability for life beyond Earth 19–21, 100–1

Index

immunity of a colony to threats on Earth   52–4, 101–2
international space station   56–8

Lagrange points   8, 105
legal issues   36–8
life already on Mars   35–6

Mars, as a site for colonization   10, 23–6
Mars One   10, 27–8, 57–61
  MIT study of the costs   59–62, 65, 114
Mars rovers, cost and weight   48
moon (of the earth)   6, 20, 26, 37, 108
moral argument   2, 118
moral hazard   55
Musk, Elon   10, 11, 18, 27–8, 36, 62, 64, 66, 68–9, 71, 85, 104, 109, 116

NASA habitat-constructing competition   31, 109
nuclear war   2–3, 17, 101, 104

opportunity costs   70–3
ozone layer   7, 69

perchlorates   43–4, 61, 111
Philae   20
political aspect to decisions   73–7, 117
Popper, Karl   107, 120
population size of a colony   42, 64–5
precautionary principle   18, 107
predicting the future   23, 98, 120

radiation   24–5, 32–3, 41, 46–52, 69, 87, 99, 101, 110, 113
Rawls, John   117
refuges for human survival   10–11, 40–4, 54, 78, 114
risk compensation   55, 113
rocket fuel   28, 30, 69
Rosetta   20

safety of a colony from threats on Earth   52–6, 101–2
seeding   91–4, 119
Sinclair, Upton   74
Singapore   65
spaceflight, not for the sake of colonization   19–21
SpaceX   110, 27–9, 62–5, 109
surrogate experience   95–8

technological society, complexity of   41–2, 44–5, 64–5, 101, 112
terraforming   33–6, 109–10
timescale for decisions   38–9, 100
threats to human survival   2–9
3D printing   31, 54, 112
Tyson, Neil Degrasse   36, 110

values versus wants   83–5, 118
Venter, Craig   93, 118
viruses   3, 8, 47–8, 52, 104
volcanism   7

Walkowicz, Lucianne   55
water on Mars   24, 31, 40–1, 52, 59–62, 70, 108
Wilson, E. O.   71